Cell Biology: Meiosis and Mitosis

Cell Biology: Meiosis and Mitosis

Edited by
Griffith Reid

Larsen & Keller
www.larsen-keller.com

Cell Biology: Meiosis and Mitosis
Edited by Griffith Reid
ISBN: 978-1-63549-182-1 (Hardback)

Larsen & Keller
Published by Larsen and Keller Education,
5 Penn Plaza,
19th Floor,
New York, NY 10001, USA

Cataloging-in-Publication Data

Cell biology : meiosis and mitosis / edited by Griffith Reid.
p. cm.
Includes bibliographical references and index.
ISBN 978-1-63549-182-1
1. Cytology. 2. Meiosis. 3. Mitosis. 4. Cell division. 5. Cell cycle.
I. Reid, Griffith.
QH581.2 .C45 2017
571.6--dc23

Printed and bound in China.

For more information regarding Larsen and Keller Education and its products, please visit the publisher's website www.larsen-keller.com

Table of Contents

Preface

This book presents the complex subject of meiosis and mitosis in the most comprehensible and easy to understand language. It elucidates the various methods and theories of these process. Meiosis and mitosis are the processes of cell division that occur in cells. It is an important part of the cell cycle. The topics included in the text are of utmost significance and bound to provide incredible insights to readers. Coherent flow of topics, student-friendly language and extensive use of examples make this an invaluable source of knowledge. The book is appropriate for those seeking detailed information in this area.

A detailed account of the significant topics covered in this book is provided below:

Chapter 1- This chapter will provide an integrated understanding of meiosis and mitosis. Meiosis is a specialized cell division, which reduces the chromosome number by half. Mitosis is the process of division of chromosomes into two new nuclei. This chapter is an overview of the subject matter incorporating all the major aspects of meiosis and mitosis.

Chapter 2- Cytokinesis, DNA replication, synapsis, mitotic index are some of the processes of meiosis and mitosis. These perform a variety of functions in a living organism. The processes elucidated in the following content are of vital importance, and provide a better understanding of meiosis and mitosis.

Chapter 3- An organism whose cells contain a nucleus within its membrane is a eukaryote. Reproduction in eukaryotes can occur both sexually and asexually. While it happens asexually in mitosis, it occurs sexually in meiosis. This chapter on meiosis seeks to explain the development of sexual reproduction and the specific evolutionary characteristics that were propagated because of the same.

Chapter 4- The errors and variations discussed in this chapter are aneuploidy, anaphase lag and endoreduplication. Aneuploidy is when a cell contains an incorrect number of chromosomes. An extra or a missing chromosome can cause a number of genetic disorders. The topics discussed in the chapter are of great importance to broaden the existing knowledge on mitosis.

Chapter 5- The series of events that take place when a cell gets divided is known as cell division. The cell cycle is also divided into three periods: interphase, the mitotic phase, and cytokinesis. During interphase, the cell grows and accumulates nutrients, preparing itself for cell division, and then during the mitotic phase, the chromosomes separate leading the cell to its final stage, the cytokinesis. This section provides a study of cell division and cell cycle.

Chapter 6- The period between mitotic divisions, known as G1 phase, S phase and G2 phase, is an interphase. Mitosis is a continuous process, which is divided into five stages which are prophase, prometaphase, metaphase, anaphase and telophase. The major categories of mitosis and cell cycle are dealt with great details in the chapter.

Chapter 7- Cells can be divided based on the function and genetic variation they display. This chapter explores the various types of cells like germ cells, somatic cells, zygote, gamete and phragmoplast. The content provides information on the differences between the various types.

Chapter 8- Homologous chromosomes are chromosomes that align perfectly with one another as they share the same genes and loci. This chapter provides the reader with an overview on homologous chromosomes, sister chromatids, the process of cohesin etc.

Chapter 9- Microtubules are an essential part of the cytoskeleton in eukaryotic cells and certain types of bacteria. This chapter studies microtubules, their functions and their intracellular organization. Also included are topics about the various types of microtubules like preprophase band, astral microtubules and phycoplast.

It gives me an immense pleasure to thank our entire team for their efforts. Finally in the end, I would like to thank my family and colleagues who have been a great source of inspiration and support.

Editor

1

Introduction to Meiosis and Mitosis

This chapter will provide an integrated understanding of meiosis and mitosis. Meiosis is a specialized cell division, which reduces the chromosome number by half. Mitosis is the process of division of chromosomes into two new nuclei. This chapter is an overview of the subject matter incorporating all the major aspects of meiosis and mitosis.

Meiosis

Meiosis is a specialized type of cell division that reduces the chromosome number by half. This process occurs in all sexually reproducing single-celled and multicellular eukaryotes, including animals, plants, and fungi. Errors in meiosis resulting in aneuploidy are the leading known cause of miscarriage and the most frequent genetic cause of developmental disabilities.

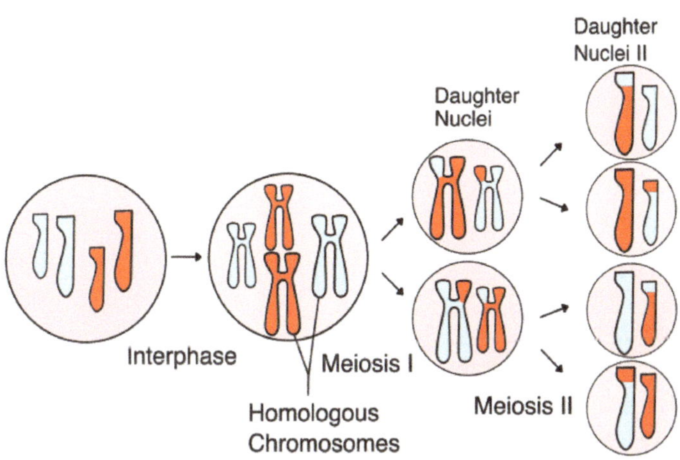

In meiosis, the chromosomes duplicate (during interphase) and homologous chromosomes exchange genetic information (chromosomal crossover) during the first division, called meiosis I. The daughter cells divide again in meiosis II, splitting up sister chromatids to form haploid gametes. Male and female gametes fuse during fertilization, creating a diploid cell with a complete set of paired chromosomes.

In meiosis, DNA replication is followed by two rounds of cell division to produce four potential daughter cells, each with half the number of chromosomes as the original parent cell. The two meiotic divisions are known as *Meiosis I* and *Meiosis II*. Before meiosis begins, during S phase of the cell cycle, the DNA of each chromosome is replicated so that it consists of two identical sister chromatids, which remain held together through sister chromatid cohesion. This S-phase can be referred to as "premeiotic

S-phase" or "meiotic S-phase." Immediately following DNA replication, meiotic cells enter a prolonged G2-like stage known as meiotic prophase. During this time, homologous chromosomes pair with each other and undergo genetic recombination, a programmed process in which DNA is cut and then repaired, which allows them to exchange some of their genetic information. A subset of recombination events results in crossovers, which create physical links known as chiasmata (singular: chiasma, for the Greek letter Chi) between the homologous chromosomes. In most organisms, these links are essential to direct each pair of homologous chromosomes to segregate away from each other during Meiosis I, resulting in two haploid cells that have half the number of chromosomes as the parent cell. During Meiosis II, the cohesion between sister chromatids is released and they segregate from one another, as during mitosis. In some cases all four of the meiotic products form gametes such as sperm, spores, or pollen. In female animals, three of the four meiotic products are typically eliminated by extrusion into polar bodies, and only one cell develops to produce an ovum.

Because the number of chromosomes is halved during meiosis, gametes can fuse (i.e. fertilization) to form a diploid zygote that contains two copies of each chromosome, one from each parent. Thus, alternating cycles of meiosis and fertilization enable sexual reproduction, with successive generations maintaining the same number of chromosomes. For example, diploid human cells contain 23 pairs of chromosomes (46 total), half of maternal origin and half of paternal origin. Meiosis produces haploid gametes (ova or sperm) that contain one set of 23 chromosomes. When two gametes (an egg and a sperm) fuse, the resulting zygote is once again diploid, with the mother and father each contributing 23 chromosomes. This same pattern, but not the same number of chromosomes, occurs in all organisms that utilize meiosis.

Overview

While the process of meiosis is related to the more general cell division process of mitosis, it differs in two important respects:

recombination	meiosis	shuffles the genes between the two chromosomes in each pair (one received from each parent), producing recombinant chromosomes with unique genetic combinations in every gamete
	mitosis	occurs only if needed to repair DNA damage; usually occurs between identical sister chromatids and does not result in genetic changes
chromosome number (ploidy)	meiosis	produces four genetically unique cells, each with half the number of chromosomes as in the parent
	mitosis	produces two genetically identical cells, each with the same number of chromosomes as in the parent

Meiosis begins with a diploid cell, which contains two copies of each chromosome, termed homologs. First, the cell undergoes DNA replication, so each homolog now consists of two identical sister chromatids. Then each set of homologs pair with each other and exchange DNA by homologous recombination leading to physical connections (crossovers) between the homologs. In the first meiotic division, the homologs are segregated to separate daughter cells by the spindle apparatus. The cells then proceed to a second division without an intervening round of DNA replication. The sister chromatids are segregated to separate daughter cells to produce a total of four haploid cells. Female animals employ a slight variation on this pattern and produce one large ovum and two small polar bodies. Because of recombination, an individual chromatid can consist of a new combination of maternal and paternal DNA, resulting in offspring that are genetically distinct from either parent. Furthermore, an individual gamete can include an assortment of maternal, paternal, and recombinant chromatids. This genetic diversity resulting from sexual reproduction contributes to the variation in traits upon which natural selection can act.

Meiosis uses many of the same mechanisms as mitosis, the type of cell division used by eukaryotes to divide one cell into two identical daughter cells. In some plants, fungi, and protists meiosis results in the formation of spores: haploid cells that can divide vegetatively without undergoing fertilization. Some eukaryotes, like bdelloid rotifers, do not have the ability to carry out meiosis and have acquired the ability to reproduce by parthenogenesis.

Meiosis does not occur in archaea or bacteria, which generally reproduce via asexual processes such as binary fission. However, a "sexual" process known as horizontal gene transfer involves the transfer of DNA from one bacterium or archaeon to another and recombination of these DNA molecules of different parental origin.

History

Meiosis was discovered and described for the first time in sea urchin eggs in 1876 by the German biologist Oscar Hertwig. It was described again in 1883, at the level of chromosomes, by the Belgian zoologist Edouard Van Beneden, in *Ascaris* roundworm eggs. The significance of meiosis for reproduction and inheritance, however, was described only in 1890 by German biologist August Weismann, who noted that two cell divisions were necessary to transform one diploid cell into four haploid cells if the number of chromosomes had to be maintained. In 1911 the American geneticist Thomas Hunt Morgan detected crossovers in meiosis in the fruit fly *Drosophila melanogaster*, which helped to establish that genetic traits are transmitted on chromosomes.

The term meiosis (originally spelled "maiosis") was introduced to biology by J.B. Farmer and J.E.S. Moore in 1905:

We propose to apply the terms Maiosis or Maiotic phase to cover the whole series of nuclear changes included in the two divisions that were designated as Heterotype and Homotype by Flemming.

Occurrence in Eukaryotic Life Cycles

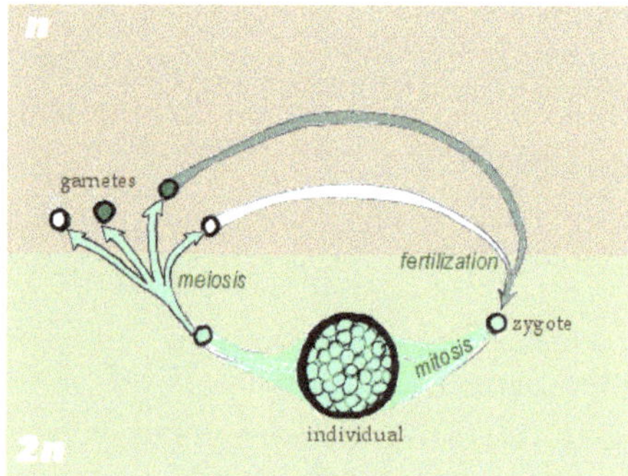

Gametic life cycle.

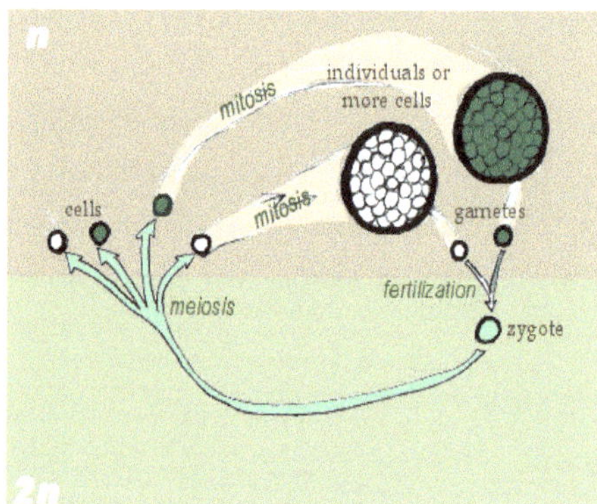

Zygotic life cycle.

Meiosis occurs in eukaryotic life cycles involving sexual reproduction, consisting of the constant cyclical process of meiosis and fertilization. This takes place alongside normal mitotic cell division. In multicellular organisms, there is an intermediary step between the diploid and haploid transition where the organism grows. At certain stages of the life cycle, germ cells produce gametes. Somatic cells make up the body of the organism and are not involved in gamete production.

Cycling meiosis and fertilization events produces a series of transitions back and forth between alternating haploid and diploid states. The organism phase of the life cycle can occur either during the diploid state (*gametic* or *diploid* life cycle), during the haploid state (*zygotic* or *haploid* life cycle), or both (*sporic* or *haplodiploid* life cycle, in which there are two distinct organism phases, one during the haploid state and the other during the diploid state). In this sense there are three types of life cycles that utilize sexual reproduction, differentiated by the location of the organism phase(s).

In the *gametic life cycle* or " diplontic life cycle", of which humans are a part, the organism is diploid, grown from a diploid cell called the zygote. The organism's diploid germ-line stem cells undergo meiosis to create haploid gametes (the spermatozoa for males and ova for females), which fertilize to form the zygote. The diploid zygote undergoes repeated cellular division by mitosis to grow into the organism.

In the *zygotic life cycle* the organism is haploid instead, spawned by the proliferation and differentiation of a single haploid cell called the gamete. Two organisms of opposing sex contribute their haploid gametes to form a diploid zygote. The zygote undergoes meiosis immediately, creating four haploid cells. These cells undergo mitosis to create the organism. Many fungi and many protozoa utilize the zygotic life cycle.

Finally, in the *sporic life cycle*, the living organism alternates between haploid and diploid states. Consequently, this cycle is also known as the alternation of generations. The diploid organism's germ-line cells undergo meiosis to produce spores. The spores proliferate by mitosis, growing into a haploid organism. The haploid organism's gamete then combines with another haploid organism's gamete, creating the zygote. The zygote undergoes repeated mitosis and differentiation to become a diploid organism again. The sporic life cycle can be considered a fusion of the gametic and zygotic life cycles.

Process

The preparatory steps that lead up to meiosis are identical in pattern and name to interphase of the mitotic cell cycle.

Interphase is divided into three phases:

- Growth 1 (G_1) phase: In this very active phase, the cell synthesizes its vast array of proteins, including the enzymes and structural proteins it will need for growth. In G_1, each of the chromosomes consists of a single linear molecule of DNA.
- Synthesis (S) phase: The genetic material is replicated; each of the cell's chromosomes duplicates to become two identical sister chromatids attached at a centromere. This replication does not change the ploidy of the cell since the centromere number remains the same. The identical sister chromatids have not yet condensed into the densely packaged chromosomes visible with the light microscope. This will take place during prophase I in meiosis.
- Growth 2 (G_2) phase: G_2 phase as seen before mitosis is not present in meiosis. Meiotic prophase corresponds most closely to the G_2 phase of the mitotic cell cycle.

Interphase is followed by meiosis I and then meiosis II. Meiosis I separates homologous chromosomes, each still made up of two sister chromatids, into two daughter cells, thus reducing the chromosome number by half. During meiosis II, sister chromatids decouple and the resultant daughter chromosomes are segregated into four daughter cells. For diploid organisms, the daughter cells resulting from meiosis are haploid and contain only one copy of each chromosome. In some species, cells enter a resting phase known as interkinesis between meiosis I and meiosis II.

Meiosis I and II are each divided into prophase, metaphase, anaphase, and telophase stages, similar in purpose to their analogous subphases in the mitotic cell cycle. Therefore, meiosis includes the stages of meiosis I (prophase I, metaphase I, anaphase I, telophase I) and meiosis II (prophase II, metaphase II, anaphase II, telophase II).

Meiosis generates gamete genetic diversity in two ways: (1) the independent orientation of homologous chromosome pairs along the metaphase plate during metaphase I and the subsequent separation of homologs during anaphase I allows a random and independent distribution of chromosomes to each daughter cell (and ultimately to gametes); and (2) physical exchange of homologous chromosomal regions by homologous recombination during prophase I results in new combinations of DNA within chromosomes.

During meiosis, specific genes are more highly transcribed. In addition to strong meiotic stage-specific expression of mRNA, there are also pervasive translational controls (e.g. selective usage of preformed mRNA), regulating the ultimate meiotic stage-specific protein expression of genes during meiosis. Thus, both transcriptional and translational controls determine the broad restructuring of meiotic cells needed to carry out meiosis.

Phases

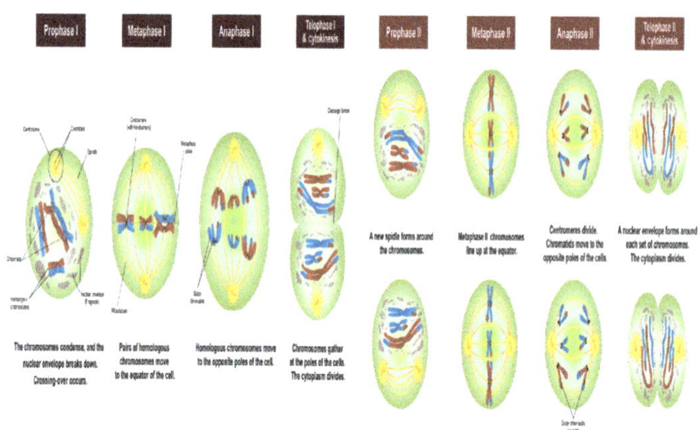

Diagram of the meiotic phases

Meiosis is divided into meiosis I and meiosis II which are further divided into Karyokinesis I and Cytokinesis I & Karyokinesis II and Cytokinesis II respectively.

Meiosis I

Meiosis I segregates homologous chromosomes, which are joined as tetrads (2n, 4c), producing two haploid cells (n chromosomes, 23 in humans) which each contain chromatid pairs (1n, 2c). Because the ploidy is reduced from diploid to haploid, meiosis I is referred to as a *reductional division*. Meiosis II is an *equational division* analogous to mitosis, in which the sister chromatids are segregated, creating four haploid daughter cells (1n, 1c).

Prophase I

Prophase I is typically the longest phase of meiosis. During prophase I, homologous chromosomes pair and exchange DNA in a process called homologous recombination. This often results in chromosomal crossover. This process is critical for pairing between homologous chromosomes and hence for accurate segregation of the chromosomes at the first meiosis division. The new combinations of DNA created during crossover are a significant source of genetic variation, and result in new combinations of alleles, which may be beneficial. The paired and replicated chromosomes are called bivalents or tetrads, which have two chromosomes and four chromatids, with one chromosome coming from each parent. The process of pairing the homologous chromosomes is called synapsis. At this stage, non-sister chromatids may cross-over at points called chiasmata (plural; singular chiasma). Prophase I has historically been divided into a series of substages which are named according to the appearance of chromosomes.

Leptotene

The first stage of prophase I is the *leptotene* stage, also known as *leptonema*, from Greek words meaning "thin threads". In this stage of prophase I, individual chromosomes—each consisting of two sister chromatids—become "individualized" to form visible strands within the nucleus. The two sister chromatids closely associated and are visually indistinguishable from one another. During leptotene, lateral elements of the synaptonemal complex assemble. Leptotene is of very short duration and progressive condensation and coiling of chromosome fibers takes place.

Zygotene

The *zygotene* stage, also known as *zygonema*, from Greek words meaning "paired threads", occurs as the chromosomes approximately line up with each other into homologous chromosome pairs. In some organisms, this is called the bouquet stage because of the way the telomeres cluster at one end of the nucleus. At this stage, the synapsis (pairing/coming together) of homologous chromosomes takes place, facilitated by assembly of central element of the synaptonemal complex. Pairing is brought about in a zipper-like fashion and may start at the centromere (procentric), at the chromosome ends (proterminal), or at any other portion (intermediate). Individuals of a pair are equal in length and in position of the centromere. Thus pairing is highly specific and exact. The paired chromosomes are called bivalent or tetrad chromosomes.

Pachytene

The *pachytene* stage, also known as *pachynema*, from Greek words meaning "thick threads". At this point a tetrad of the chromosomes has formed known as a bivalent. This is the stage when chromosomal crossover (crossing over) occurs. Nonsister chromatids of homologous chromosomes may exchange segments over regions of homology. Sex chromosomes, however, are not wholly identical, and only exchange information over a small region of homology. At the sites where exchange happens, chiasmata form. The exchange of information between the non-sister chromatids results in a recombination of information; each chromosome has the complete set of information it had before, and there are no gaps formed as a result of the process. Because the chromosomes cannot be distinguished in the synaptonemal complex, the actual act of crossing over is not perceivable through the microscope, and chiasmata are not visible until the next stage.

Diplotene

During the *diplotene* stage, also known as *diplonema*, from Greek words meaning "two threads", the synaptonemal complex degrades and homologous chromosomes separate from one another a little. The chromosomes themselves uncoil a bit, allowing some transcription of DNA. However, the homologous chromosomes of each bivalent remain tightly bound at chiasmata, the regions where crossing-over occurred. The chiasmata remain on the chromosomes until they are severed at the transition to anaphase I.

In mammalian and human fetal oogenesis all developing oocytes develop to this stage and are arrested before birth. This suspended state is referred to as the *dictyotene stage* or dictyate. It lasts until meiosis is resumed to prepare the oocyte for ovulation, which happens at puberty or even later.

Diakinesis

Chromosomes condense further during the *diakinesis* stage, from Greek words meaning "moving through". This is the first point in meiosis where the four parts of the tetrads are actually visible. Sites of crossing over entangle together, effectively overlapping, making chiasmata clearly visible. Other than this observation, the rest of the stage closely resembles prometaphase of mitosis; the nucleoli disappear, the nuclear membrane disintegrates into vesicles, and the meiotic spindle begins to form.

Synchronous Processes

During these stages, two centrosomes, containing a pair of centrioles in animal cells, migrate to the two poles of the cell. These centrosomes, which were duplicated during S-phase, function as microtubule organizing centers nucleating microtubules, which are essentially cellular ropes and poles. The microtubules invade the nuclear region after the nuclear envelope disintegrates, attaching to the chromosomes at the kinetochore. The kinetochore functions as a motor, pulling the chromosome along the attached microtubule toward the originating centrosome, like a train on a track. There are four kinetochores on each tetrad, but the pair of kinetochores on each sister chromatid fuses and functions as a unit during meiosis I.

Microtubules that attach to the kinetochores are known as *kinetochore microtubules*. Other microtubules will interact with microtubules from the opposite centrosome: these are called *nonkinetochore microtubules* or *polar microtubules*. A third type of microtubules, the aster microtubules, radiates from the centrosome into the cytoplasm or contacts components of the membrane skeleton.

Metaphase I

Homologous pairs move together along the metaphase plate: As *kinetochore microtubules* from both centrosomes attach to their respective kinetochores, the paired homologous chromosomes align along an equatorial plane that bisects the spindle, due to continuous counterbalancing forces exerted on the bivalents by the microtubules emanating from the two kinetochores of homologous chromosomes. The physical basis of the independent assortment of chromosomes is the random orientation of each bivalent along the metaphase plate, with respect to the orientation of the other bivalents along the same equatorial line. The protein complex cohesin holds sister chromatids together from the time of their replication until anaphase. In mitosis, the force of kinetochore microtubules pulling in opposite directions creates tension. The cell senses this tension and does not progress with anaphase until all the chromosomes are properly bi-oriented. In meiosis, establishing tension requires at least one crossover per chromosome pair in addition to cohesin between sister chromatids.

Anaphase I

Kinetochore microtubules shorten, pulling homologous chromosomes (which consist of a pair of sister chromatids) to opposite poles. Nonkinetochore microtubules lengthen, pushing the centrosomes farther apart. The cell elongates in preparation for division down the center. Unlike in mitosis, only the cohesin from the chromosome arms is degraded while the cohesin surrounding the centromere remains protected. This allows the sister chromatids to remain together while homologs are segregated.

Telophase I

The first meiotic division effectively ends when the chromosomes arrive at the poles. Each daughter cell now has half the number of chromosomes but each chromosome consists of a pair of chromatids. The microtubules that make up the spindle network disappear, and a new nuclear membrane surrounds each haploid set. The chromosomes uncoil back into chromatin. Cytokinesis, the pinching of the cell membrane in animal cells or the formation of the cell wall in plant cells, occurs, completing the creation of two daughter cells. Sister chromatids remain attached during telophase I.

Cells may enter a period of rest known as interkinesis or interphase II. No DNA replication occurs during this stage.

Meiosis II

Meiosis II is the second meiotic division, and usually involves equational segregation, or separation of sister chromatids. Mechanically, the process is similar to mitosis, though its genetic results are fundamentally different. The end result is production of four haploid cells (n chromosomes, 23 in humans) from the two haploid cells (with n chromosomes, each consisting of two sister chromatids) produced in meiosis I. The four main steps of Meiosis II are: Prophase II, Metaphase II, Anaphase II, and Telophase II.

In prophase II we see the disappearance of the nucleoli and the nuclear envelope again as well as the shortening and thickening of the chromatids. Centrosomes move to the polar regions and arrange spindle fibers for the second meiotic division.

In metaphase II, the centromeres contain two kinetochores that attach to spindle fibers from the centrosomes at opposite poles. The new equatorial metaphase plate is rotated by 90 degrees when compared to meiosis I, perpendicular to the previous plate.

This is followed by anaphase II, in which the remaining centromeric cohesin is cleaved allowing the sister chromatids to segregate. The sister chromatids by convention are now called sister chromosomes as they move toward opposing poles.

The process ends with telophase II, which is similar to telophase I, and is marked by decondensation and lengthening of the chromosomes and the disassembly of the spindle. Nuclear envelopes reform and cleavage or cell plate formation eventually produces a total of four daughter cells, each with a haploid set of chromosomes.

Meiosis is now complete and ends up with four new daughter cells.

Origin and Function

The origin and function of meiosis are fundamental to understanding the evolution of sexual reproduction in Eukaryotes. There is no current consensus among biologists on the questions of how sex in Eukaryotes arose in evolution, what basic function sexual reproduction serves, and why it is maintained, given the basic two-fold cost of sex. It is clear that it evolved over 1.2 billion years ago, and that almost all species which are descendents of the original sexually reproducing species are still sexual reproducers, including plants, fungi, and animals.

Meiosis is ubiquitous among eukaryotes. It occurs in single-celled organisms such as yeast, as well as in multicellular organisms, such as humans. Eukaryotes arose from prokaryotes more than 1.5 billion years ago, and the earliest eukaryotes were likely single-celled organisms. To understand sex in eukaryotes, it is necessary to understand (1) how meiosis arose in single celled eukaryotes, and (2) the function of meiosis.

Nondisjunction

The normal separation of chromosomes in meiosis I or sister chromatids in meiosis II is termed *disjunction*. When the segregation is not normal, it is called *nondisjunction*. This results in the production of gametes which have either too many or too few of a particular chromosome, and is a common mechanism for trisomy or monosomy. Nondisjunction can occur in the meiosis I or meiosis II, phases of cellular reproduction, or during mitosis.

Most monosomic and trisomic human embryos are not viable, but some aneuploidies can be tolerated, such as trisomy for the smallest chromosome, chromosome 21. Phenotypes of these aneuploidies range from severe developmental disorders to asymptomatic. Medical conditions include but are not limited to:

- Down Syndrome - trisomy of chromosome 21
- Patau Syndrome - trisomy of chromosome 13
- Edward Syndrome - trisomy of chromosome 18
- Klinefelter Syndrome - extra X chromosomes in males - i.e. XXY, XXXY, XXXXY, etc.
- Turner Syndrome - lacking of one X chromosome in females - i.e. Xo
- Triple X syndrome - an extra X chromosome in females
- XYY Syndrome - an extra Y chromosome in males.

The probability of nondisjunction in human oocytes increases with increasing maternal age, presumably due to loss of cohesin over time.

Meiosis in Plants and Animals

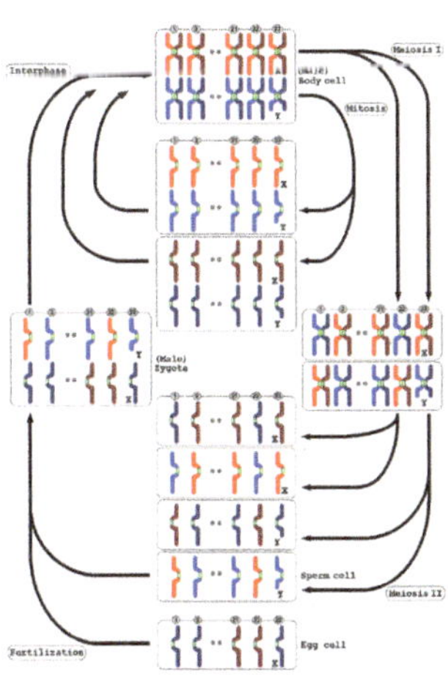

Overview of chromatides' and chromosomes' distribution within the mitotic and meiotic cycle of a male human cell

Meiosis occurs in all animals and plants. The end result, the production of gametes with half the number of chromosomes as the parent cell, is the same, but the detailed process is different. In animals, meiosis produces gametes directly. In land plants and some algae, there is an alternation of generations such that meiosis in the diploid sporophyte generation produces haploid spores. These spores multiply by mitosis, developing into the haploid gametophyte generation, which then gives rise to gametes directly (i.e. without further meiosis). In both animals and plants, the final stage is for the gametes to fuse, restoring the original number of chromosomes.

Meiosis in Mammals

In females, meiosis occurs in cells known as oocytes (singular: oocyte). Each oocyte that initiates meiosis divides twice, unequally in each case. The first division produces a daughter cell that will undergo a second division, and a much smaller "polar body" that is extruded from the surface of the cell and does not divide further. Following Meiosis II, a "second polar body" is extruded, and the single remaining haploid cell enlarges to become an ovum. Since the first polar body normally disintegrates rather than dividing again, meiosis in female mammals results in three products, the oocyte and two polar bodies. However, before these divisions occur, these cells stop at the diplotene stage of meiosis I and lie dormant within a protective shell of somatic cells called the follicle. Follicles begin growth at a steady pace in a process known as folliculogenesis, and a small number enter the menstrual cycle. Menstruated oocytes continue meiosis I and arrest at meiosis II until fertilization. The process of meiosis in females occurs during oogenesis, and differs from the typical meiosis in that it features a long period of meiotic arrest known as the dictyate stage and lacks the assistance of centrosomes.

In males, meiosis occurs during spermatogenesis in the seminiferous tubules of the testicles. Meiosis during spermatogenesis is specific to a type of cell called spermatocytes that will later mature to become spermatozoa.

In female mammals, meiosis begins immediately after primordial germ cells migrate to the ovary in the embryo, but in the males, meiosis begins later, at the time of puberty. It is retinoic acid, derived from the primitive kidney (mesonephros) that stimulates meiosis in ovarian oogonia. Tissues of the male testis suppress meiosis by degrading retinoic acid, a stimulator of meiosis. This is overcome at puberty when cells within seminiferous tubules called Sertoli cells start making their own retinoic acid. Sensitivity to retinoic acid is also adjusted by proteins called nanos and DAZL.

Meiosis Vs. Mitosis

In order to understand meiosis, a comparison to mitosis is helpful. The table below shows the differences between meiosis and mitosis.

	Meiosis	Mitosis
End result	Normally four cells, each with half the number of chromosomes as the parent	Two cells, having the same number of chromosomes as the parent
Function	Sexual reproduction, production of gametes (sex cells)	Cellular reproduction, growth, repair, asexual reproduction
Where does it happen?	Reproductive cells of almost all eukaryotes (animals, plants, fungi, and protists)	All proliferating cells in all eukaryotes
Steps	Prophase I, Metaphase I, Anaphase I, Telophase I, Prophase II, Metaphase II, Anaphase II, Telophase II	Prophase, Prometaphase, Metaphase, Anaphase, Telophase
Genetically same as parent?	No	Yes
Crossing over happens?	Yes, normally occurs between each pair of homologous chromosomes	Very rarely
Pairing of homologous chromosomes?	Yes	No
Cytokinesis	Occurs in Telophase I and Telophase II	Occurs in Telophase
Centromeres split	Does not occur in Anaphase I, but occurs in Anaphase II	Occurs in Anaphase

Mitosis

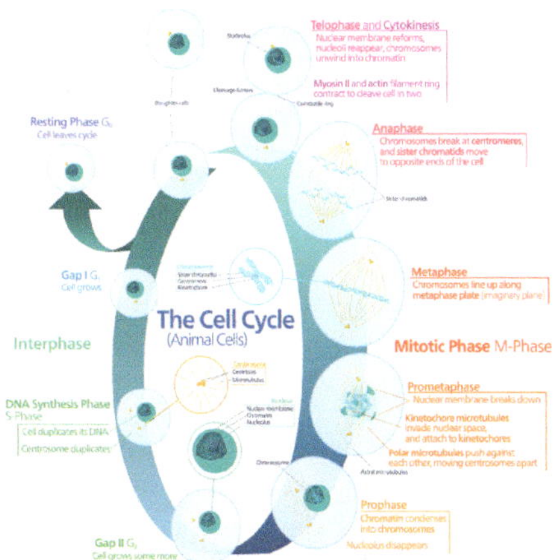

Mitosis in an animal cell (phases ordered counter-clockwise).

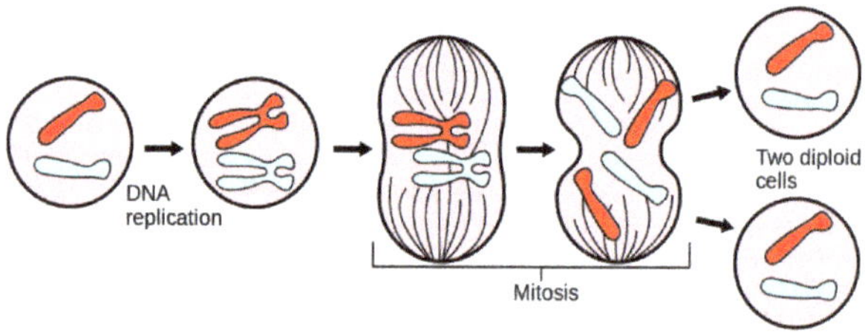

Mitosis divides the chromosomes in a cell nucleus.

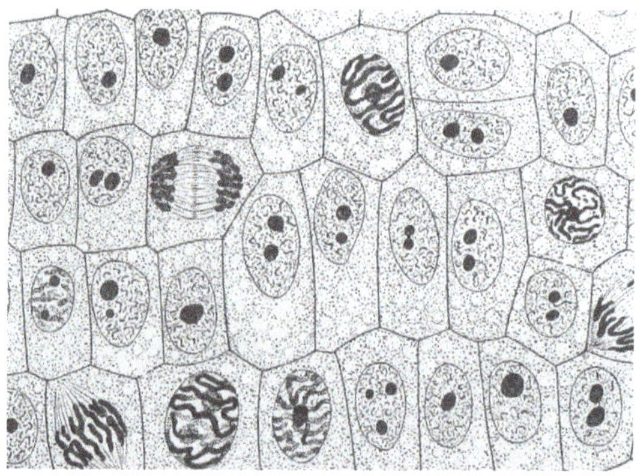

Onion (Allium) cells in different phases of the cell cycle enlarged 800 diameters.

a. non-dividing cells

b. nuclei preparing for division (spireme-stage)

c. dividing cells showing mitotic figures

e. pair of daughter-cells shortly after division

In cell biology, Mitosis is a part of the cell cycle in which chromosomes in a cell nucleus are separated into two identical sets of chromosomes, and each set ends up in its own nucleus. In general, mitosis (division of the nucleus) is preceded by the S stage of interphase (during which the DNA is replicated) and is often accompanied or followed by cytokinesis, which divides the cytoplasm, organelles and cell membrane into two new cells containing roughly equal shares of these cellular components. Mitosis and cytokinesis together define the mitotic (M) phase of an animal cell cycle—the division of the mother cell into two daughter cells, genetically identical to each other and to their parent cell.

The process of mitosis is divided into stages corresponding to the completion of one set of activities and the start of the next. These stages are prophase, prometaphase, metaphase, anaphase, and telophase. During mitosis, the chromosomes, which have already duplicated, condense and attach to spindle fibers that pull one copy of each chromosome to opposite sides of the cell. The result is two genetically identical daughter nuclei. The rest of the cell may then continue to divide by cytokinesis to produce two daughter cells. Producing three or more daughter cells instead of normal two is a mitotic error called tripolar mitosis or multipolar mitosis (direct cell triplication / multiplication). Other errors during mitosis can induce apoptosis (programmed cell death) or cause mutations. Certain types of cancer can arise from such mutations.

Mitosis occurs only in eukaryotic cells and the process varies in different organisms. For example, animals undergo an "open" mitosis, where the nuclear envelope breaks down before the chromosomes separate, while fungi undergo a "closed" mitosis, where chromosomes divide within an intact cell nucleus. Furthermore, most animal cells undergo a shape change, known as mitotic cell rounding, to adopt a near spherical morphology at the start of mitosis. Prokaryotic cells, which lack a nucleus, divide by a different process called binary fission.

Discovery

German zoologist Otto Bütschli might have claimed the discovery of the process presently known as "mitosis", a term coined by Walther Flemming in 1882.

Mitosis was discovered in frog, rabbit, and cat cornea cells in 1873 and described for the first time by the Polish histologist Wacław Mayzel in 1875. The term is derived from the Greek word μίτος *mitos* "warp thread".

Overview of Mitosis

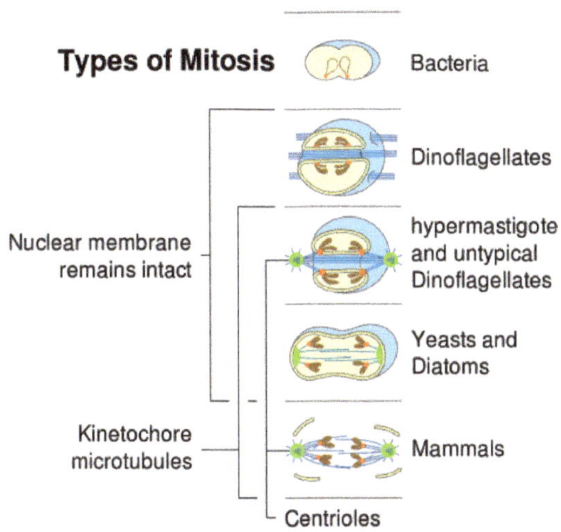

Types of mitosis

The primary result of mitosis and cytokinesis is the transfer of a parent cell's genome into two daughter cells. The genome is composed of a number of chromosomes—complexes of tightly coiled DNA that contain genetic information vital for proper cell function. Because each resultant daughter cell should be genetically identical to the parent cell, the parent cell must make a copy of each chromosome before mitosis. This occurs during the S phase of interphase. Chromosome duplication results in two identical *sister chromatids* bound together by cohesin proteins at the *centromere*.

When mitosis begins, the chromosomes condense and become visible. In some eukaryotes, for example animals, the nuclear envelope, which segregates the DNA from the cytoplasm, disintegrates into small vesicles. The nucleolus, which makes ribosomes in the cell, also disappears. Microtubules project from opposite ends of the cell, attach to the centromeres, and align the chromosomes centrally within the cell. The microtubules then contract to pull the sister chromatids of each chromosome apart. Sister chromatids at this point are called *daughter chromosomes*. As the cell elongates, corresponding daughter chromosomes are pulled toward opposite ends of the cell and condense maximally in late anaphase. A new nuclear envelope forms around the separated daughter chromosomes, which decondense to form interphase nuclei.

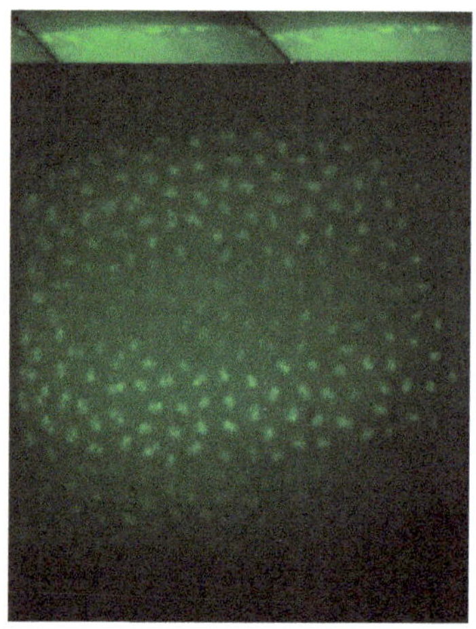

Time-lapse video of mitosis in a *Drosophila melanogaster* embryo.

During mitotic progression, typically after anaphase onset, the cell may undergo cytokinesis. In animal cells, a cell membrane pinches inward between the two developing nuclei to produce two new cells. In plant cells, a cell plate forms between the two nuclei. Cytokinesis does not always occur; coenocytic (a type of multinucleate condition) cells undergo mitosis without cytokinesis.

Phases of Cell Cycle and Mitosis

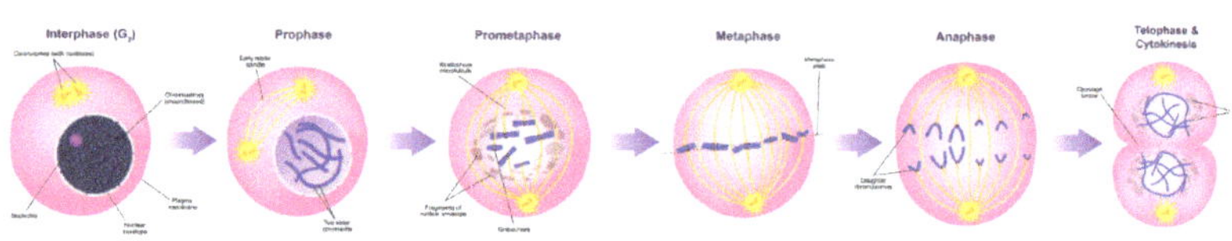

Interphase

The mitotic phase is a relatively short period of the cell cycle. It alternates with the much longer *interphase*, where the cell prepares itself for the process of cell division. Interphase is divided into three phases: G_1 (first gap), S (synthesis), and G_2 (second gap). During all three phases, the cell grows by producing proteins and cytoplasmic organelles. However, chromosomes are replicated only during the S phase. Thus, a cell grows (G_1), continues to grow as it duplicates its chromosomes (S), grows more and prepares for mitosis (G_2), and finally divides (M) before restarting the cycle. All these phases in the cell cycle are highly regulated by cyclins, cyclin-dependent kinases, and other cell cycle proteins. The phases follow one another in strict order and there are "checkpoints" that give the cell cues to proceed from one phase to another. Cells may also temporarily or permanently leave the cell cycle and enter G_0 phase to stop dividing. This can occur when cells become overcrowded (density-dependent inhibition) or when they differentiate to carry out specific functions for the organism, as is the case for human heart muscle cells and neurons. Some G_0 cells have the ability to re-enter the cell cycle.

Preprophase (Plant Cells)

In plant cells only, prophase is preceded by a pre-prophase stage. In highly vacuolated plant cells, the nucleus has to migrate into the center of the cell before mitosis can begin. This is achieved through the formation of a phragmosome, a transverse sheet of cytoplasm that bisects the cell along the future plane of cell division. In addition to phragmosome formation, preprophase is characterized by the formation of a ring of microtubules and actin filaments (called preprophase band) underneath the plasma membrane around the equatorial plane of the future mitotic spindle. This band marks the position where the cell will eventually divide. The cells of higher plants (such as the flowering plants) lack centrioles; instead, microtubules form a spindle on the surface of the nucleus and are then organized into a spindle by the chromosomes themselves, after the nuclear envelope breaks down. The preprophase band disappears during nuclear envelope breakdown and spindle formation in prometaphase.

Prophase

During prophase, which occurs after G_2 interphase, the cell prepares to divide by tightly condensing its chromosomes and initiating mitotic spindle formation, this process is called chromosome condensation. During interphase, the genetic material in the nucleus consists of loosely packed chromatin. At the onset of prophase, chromatin fibers condense into discrete chromosomes that are typically visible at high magnification through a light microscope. In this stage, chromosomes are long, thin and thread-like. Each chromosome has two chromatids. The two chromatids are joined at a place called centromere.

Gene transcription ceases during prophase and does not resume until late anaphase to early G1 phase. The nucleolus also disappears during early prophase.

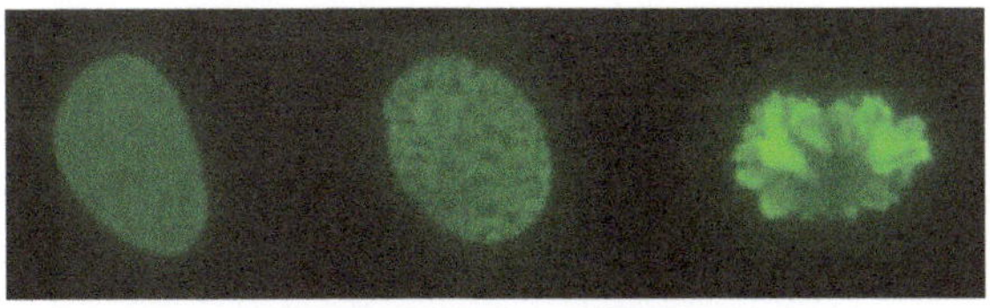

Condensing chromosomes. Interphase nucleus (left), condensing chromosomes (middle) and condensed chromosomes (right).

Close to the nucleus of animal cells are structures called centrosomes, consisting of a pair of centrioles surrounded by a loose collection of proteins. The centrosome is the coordinating center for the cell's microtubules. A cell inherits a single centrosome at cell division, which is duplicated by the cell before a new round of mitosis begins, giving a pair of centrosomes. The two centrosomes polymerize tubulin to help form a microtubule spindle apparatus. Motor proteins then push the centrosomes along these microtubules to opposite sides of the cell. Although centrosomes help organize microtubule assembly, they are not essential for the formation of the spindle apparatus, since they are absent from plants, and are not absolutely required for animal cell mitosis.

Prometaphase

At the beginning of prometaphase in animal cells, phosphorylation of nuclear lamins causes the nuclear envelope to disintegrate into small membrane vesicles. As this happens, microtubules invade the nuclear space. This is called *open mitosis*, and it occurs in some multicellular organisms. Fungi and some protists, such as algae or trichomonads, undergo a variation called *closed mitosis* where the spindle forms inside the nucleus, or the microtubules penetrate the intact nuclear envelope.

In late prometaphase, *kinetochore microtubules* begin to search for and attach to chromosomal kinetochores. A *kinetochore* is a proteinaceous microtubule-binding structure that forms on the chromosomal centromere during late prophase. A number of *polar microtubules* find and interact with corresponding polar microtubules from the opposite centrosome to form the mitotic spindle. Although the kinetochore structure and function are not fully understood, it is known that it contains some form of molecular motor. When a microtubule connects with the kinetochore, the motor activates, using energy from ATP to "crawl" up the tube toward the originating centrosome. This motor activity, coupled with polymerisation and depolymerisation of microtubules, provides the pulling force necessary to later separate the chromosome's two chromatids.

Metaphase

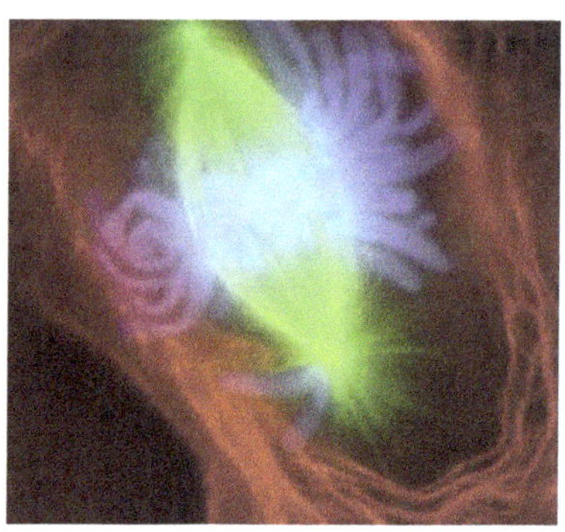

A cell in late metaphase. All chromosomes (blue) but one have arrived at the metaphase plate.

After the microtubules have located and attached to the kinetochores in prometaphase, the two centrosomes begin pulling the chromosomes towards opposite ends of the cell. The resulting tension causes the chromosomes to align along the *metaphase plate* or *equatorial plane*, an imaginary line that is centrally located between the two centrosomes (at approximately the midline of the cell). To ensure equitable distribution of chromosomes at the end of mitosis, the *metaphase checkpoint* guarantees that kinetochores are properly attached to the mitotic spindle and that the chromosomes are aligned along the metaphase plate. If the cell successfully passes through the metaphase checkpoint, it proceeds to anaphase.

Anaphase

During *anaphase A*, the cohesins that bind sister chromatids together are cleaved, forming two identical daughter chromosomes. Shortening of the kinetochore microtubules pulls the newly formed daughter chromosomes to opposite ends of the cell. During *anaphase B*, polar microtubules push against each other, causing the cell to elongate. In late anaphase, chromosomes also reach their overall maximal condensation level, to help chromosome segregation and the re-formation of the nucleus. In most animal cells, anaphase A precedes anaphase B, but some vertebrate egg cells demonstrate the opposite order of events.

Telophase

Telophase is a reversal of prophase and prometaphase events. At telophase, the polar microtubules continue to lengthen, elongating the cell even more. If the nuclear envelope has broken down, a new nuclear envelope forms using the membrane vesicles of the parent cell's old nuclear envelope. The new envelope forms around each set of separated daughter chromosomes (though the membrane does not enclose the centrosomes) and the nucleolus reappears. Both sets of chromosomes, now surrounded by new nuclear membrane, begin to "relax" or decondense. Mitosis is complete. Each daughter nucleus has an identical set of chromosomes. Cell division may or may not occur at this time depending on the organism.

Cytokinesis

Cytokinesis is not a phase of mitosis but rather a separate process, necessary for completing cell division. In animal cells, a cleavage furrow (pinch) containing a contractile ring develops where the metaphase plate used to be, pinching off the separated nuclei. In both animal and plant cells, cell division is also driven by vesicles derived from the Golgi apparatus, which move along microtubules to the middle of the cell. In plants, this structure coalesces into a cell plate at the center of the phragmoplast and develops into a cell wall, separating the two nuclei. The phragmoplast is a microtubule structure typical for higher plants, whereas some green algae use a phycoplast microtubule array during cytokinesis. Each daughter cell has a complete copy of the genome of its parent cell. The end of cytokinesis marks the end of the M-phase.

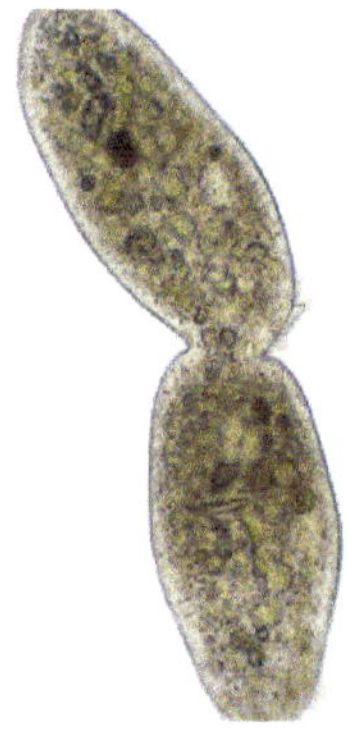

Cilliate undergoing cytokinesis, with the cleavage furrow being clearly visible

There are many cells where mitosis and cytokinesis occur separately, forming single cells with multiple nuclei. The most notable occurrence of this is among the fungi, slime molds, and coenocytic algae, but the phenomenon is found in various other organisms. Even in animals, cytokinesis and mitosis may occur independently, for instance during certain stages of fruit fly embryonic development.

Significance

Mitosis is important for the maintenance of the chromosomal set; each cell formed receives chromosomes that are alike in composition and equal in number to the chromosomes of the parent cell.

Mitosis occurs in the following circumstances:

Development and growth

> The number of cells within an organism increases by mitosis. This is the basis of the development of a multicellular body from a single cell, i.e., zygote and also the basis of the growth of a multicellular body.

Cell replacement

> In some parts of body, e.g. skin and digestive tract, cells are constantly sloughed off and replaced by new ones. New cells are formed by mitosis and so are exact copies of the cells being replaced. In like manner, red blood cells have short lifespan (only about 4 months) and new RBCs are formed by mitosis.

Regeneration

> Some organisms can regenerate body parts. The production of new cells in such instances is achieved by mitosis. For example, starfish regenerate lost arms through mitosis.

Asexual reproduction

Some organisms produce genetically similar offspring through asexual reproduction. For example, the hydra reproduces asexually by budding. The cells at the surface of hydra undergo mitosis and form a mass called a bud. Mitosis continues in the cells of the bud and this grows into a new individual. The same division happens during asexual reproduction or vegetative propagation in plants.

Cell Shape Changes During Mitosis

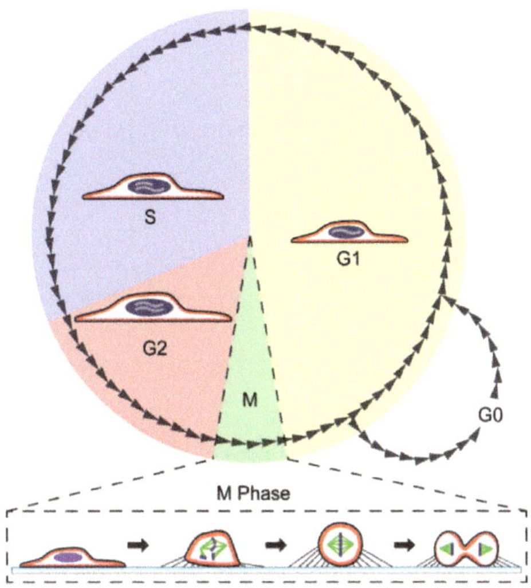

Cell shape changes through mitosis for a typical animal cell cultured on a flat surface. The cell undergoes mitotic cell rounding during spindle assembly and then divides via cytokinesis. The actomyosin cortex is depicted in red, DNA/chromosomes purple, microtubules green, and membrane and retraction fibers in black. Rounding also occurs in live tissue, as described in the text.

In animal tissue, most cells round up to a near-spherical shape during mitosis. In epithelia and epidermis, an efficient rounding process is correlated with proper mitotic spindle alignment and subsequent correct positioning of daughter cells. Moreover, researchers have found that if rounding is heavily suppressed it may result in spindle defects, primarily pole splitting and failure to efficiently capture chromosomes. Therefore, mitotic cell rounding is thought to play a protective role in ensuring accurate mitosis.

Rounding forces are driven by reorganization of F-actin and myosin (actomyosin) into a contractile homogeneous cell cortex that 1) rigidifies the cell periphery and 2) facilitates generation of intracellular hydrostatic pressure (up to 10 fold higher than interphase). The generation of intracellular pressure is particularly critical under confinement, such as would be important in a tissue scenario, where outward forces must be produced to round up against surrounding cells and/or the extracellular matrix. Generation of pressure is dependent on formin-mediated F-actin nucleation and Rho kinase (ROCK)-mediated myosin II contraction, both of which are governed upstream by signaling pathways RhoA and ECT2 through the activity of Cdk1. Due to its importance in mitosis, the molecular components and dynamics of the mitotic actomyosin cortex is an area of active research.

Errors and Variations of Mitosis

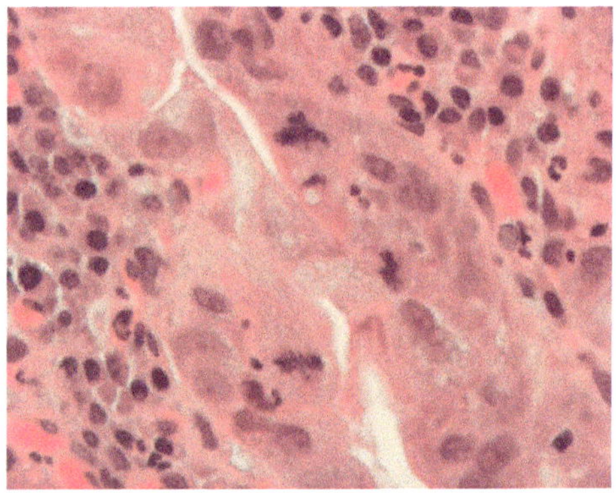

An abnormal (tripolar) mitosis (12 o'clock position) in a precancerous lesion of the stomach (H&E stain)

Errors can occur during mitosis, especially during early embryonic development in humans. Mitotic errors can create aneuploid cells that have too few or too many of one or more chromosomes, a condition associated with cancer. Early human embryos, cancer cells, infected or intoxicated cells can also suffer from pathological division into three or more daughter cells (tripolar or multipolar mitosis), resulting in severe errors in their chromosomal complements.

In *nondisjunction*, sister chromatids fail to separate during anaphase. One daughter cell receives both sister chromatids from the nondisjoining chromosome and the other cell receives none. As a result, the former cell gets three copies of the chromosome, a condition known as *trisomy*, and the latter will have only one copy, a condition known as *monosomy*. On occasion, when cells experience nondisjunction, they fail to complete cytokinesis and retain both nuclei in one cell, resulting in binucleated cells.

Anaphase lag occurs when the movement of one chromatid is impeded during anaphase. This may be caused by a failure of the mitotic spindle to properly attach to the chromosome. The lagging chromatid is excluded from both nuclei and is lost. Therefore, one of the daughter cells will be monosomic for that chromosome.

Endoreduplication (or endoreplication) occurs when chromosomes duplicate but the cell does not subsequently divide. This results in polyploid cells or, if the chromosomes duplicates repeatedly, polytene chromosomes. Endoreduplication is found in many species and appears to be a normal part of development. Endomitosis is a variant of endoreduplication in which cells replicate their chromosomes during S phase and enter, but prematurely terminate, mitosis. Instead of being divided into two new daughter nuclei, the replicated chromosomes are retained within the original nucleus. The cells then re-enter G_1 and S phase and replicate their chromosomes again. This may occur multiple times, increasing the chromosome number with each round of replication and endomitosis. Platelet-producing megakaryocytes go through endomitosis during cell differentiation.

References

- Maton, A.; Hopkins, J. J.; LaHart, S; Quon Warner, D.; Wright, M.; Jill, D. (1997). Cells: Building Blocks of Life. New Jersey: Prentice Hall. pp. 70–4. ISBN 0-13-423476-6.
- Olson, Mark O. J. (2011). The Nucleolus. Volume 15 of Protein Reviews. Berlin: Springer Science & Business Media. p. 15. ISBN 9781461405146.
- Iourov, I.Y.; Vorsanova, S.G.; Yurov, Y.B. (2006). "Chromosomal Variations in Mammalian Neuronal Cells: Known Facts and Attractive Hypotheses". In Jeon, Kwang J. International Review Of Cytology: A Survey of Cell Biology. 249. Waltham, MA: Academic Press. p. 146. ISBN 9780080463506.

2

Processes of Meiosis and Mitosis

Cytokinesis, DNA replication, synapsis, mitotic index are some of the processes of meiosis and mitosis. These perform a variety of functions in a living organism. The processes elucidated in the following content are of vital importance, and provide a better understanding of meiosis and mitosis.

Cytokinesis

Cytokinesis is the process during cell division in which the cytoplasm of a single eukaryotic cell is divided to form two daughter cells. It usually initiates during the late stages of mitosis, and sometimes meiosis, splitting a mitotic cell in two, to ensure that chromosome number is maintained from one generation to the next. After cytokinesis two (daughter) cells will be formed that are exact copies of the (parent) original cell. It is formed after cytokinesis, each daughter cell is in the interphase portion of the cell cycle. In animal cells, one notable exception to the normal process of cytokinesis is oogenesis (the creation of an ovum in the ovarian follicle of the ovary), where the ovum takes almost all the cytoplasm and organelles, leaving very little for the resulting polar bodies, which then die. Another form of mitosis without cytokinesis occurs in the liver, yielding multinucleate cells. In plant cells, a dividing structure known as the cell plate forms within the centre of the cytoplasm and a new cell wall forms between the two daughter cells.

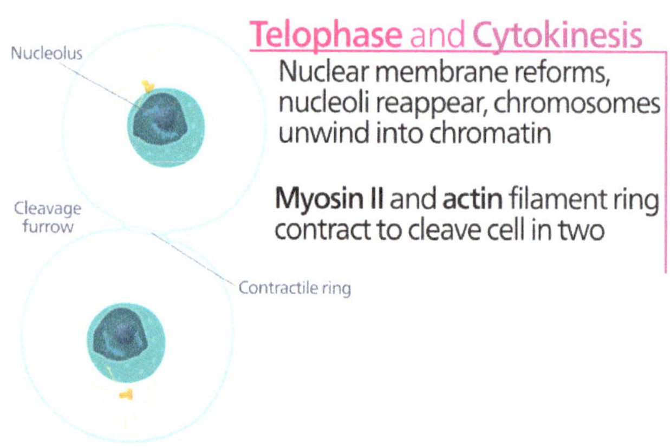

Animal cell telophase and cytokinesis

Cytokinesis is distinguished from the prokaryotic process of binary fission.

Animal Cell Cytokinesis

Contractile Ring Positioning

Animal cell cytokinesis begins shortly after the onset of sister chromatid separation in the anaphase of mitosis. A contractile ring, made of non-muscle myosin II and actin filaments, assembles equatorially (in the middle of the cell) at the cell cortex (adjacent to the cell membrane). Myosin II uses the free energy released when ATP is hydrolysed to move along these actin filaments, constricting the cell membrane to form a cleavage furrow. Continued hydrolysis causes this cleavage furrow to ingress (move inwards), a striking process that is clearly visible through a light microscope. Ingression continues until a so-called midbody structure (composed of electron-dense, proteinaceous material) is formed and the process of abscission then physically cleaves this midbody into two. Abscission depends on septin filaments beneath the cleavage furrow, which provide a structural basis to ensure the completion of cytokinesis. After cytokinesis, non-kinetochore microtubules reorganize and disappear into a new cytoskeleton as the cell cycle returns to interphase.

The position at which the contractile ring assembles is dictated by the mitotic spindle. This seems to depend upon the GTPase RhoA, which influences several downstream effectors (such as the protein kinases ROCK and citron) to promote myosin activation (by influencing the phosphorylation of Myosin regulatory light chain (rMLC)) and actin filament assembly (by regulating formin protein) at a particular region of the cell cortex.

Simultaneous with contractile ring assembly during prophase, a microtubule based structure termed the central spindle (or *spindle midzone*) forms when non-kinetochore microtubule fibres are bundled between the spindle poles. A number of different species including *H. sapiens*, *D. melanogaster* and *C. elegans* require the central spindle in order to efficiently undergo cytokinesis, although the specific phenotype described when it is absent varies from one species to the next (for example, certain Drosophila cell types are incapable of forming a cleavage furrow without the central spindle, whereas in both *C. elegans* embryos and human tissue culture cells a cleavage furrow is observed to form and ingress, but then regress before cytokinesis is complete). Seemingly vital for the formation of the central spindle (and therefore efficient cytokinesis) is a heterotetrameric protein complex called centralspindlin. Along with associated factors (such as SPD-1 in *C. elegans),* centralspindlin plays a role in bundling microtubules to form the spindle midzone during anaphase.

Timing Cytokinesis

Cytokinesis must be temporally controlled to ensure that it occurs only after sister chromatids separate during the anaphase portion of normal proliferative cell divisions. To achieve this, many components of the cytokinesis machinery are highly regulated to ensure that they are able to perform a particular function at only a particular stage of the cell cycle. Cytokinesis happens only after APC binds with CDC20. This allows for the separation of chromosomes and myosin to work simultaneously.

Plant Cell Cytokinesis

Due to the presence of a cell wall, cytokinesis in plant cells is significantly different from that in animal cells, Rather than forming a contractile ring, plant cells construct a cell plate in the middle of the cell. The stages of cell plate formation include (1) creation of the phragmoplast, an array of microtubules that guides and supports the formation of the cell plate; (2) trafficking of vesicles to the division plane and their fusion to generate a tubular-vesicular network; (3) continued fusion of membrane tubules and their transformation into membrane sheets upon the deposition of callose, followed by deposition of cellulose and other cell wall components; (4) recycling of excess membrane and other material from the cell plate; and (5) fusion with the parental cell wall

The phragmoplast is assembled from the remnants of the mitotic spindle, and serves as a track for the trafficking of vesicles to the phragmoplast midzone. These vesicles contain lipids, proteins and carbohydrates needed for the formation of a new cell boundary. Electron tomographic studies have identified the Golgi apparatus as the source of these vesicles, but other studies have suggested that they contain endocytosed material as well.

These tubules then widen and fuse laterally with each other, eventually forming a planar, fenestrated sheet . As the cell plate matures, large amounts of membrane material are removed via clathrin-mediated endocytosis Eventually, the edges of the cell plate fuse with the parental plasma membrane, often in an asymmetrical fashion, thus completing cytokinesis. The remaining fenestrae contain strands of endoplasmic reticulum passing through them, and are thought to be the precursors of plasmodesmata .

The construction of the new cell wall begins within the lumen of the narrow tubules of the young cell plate. The order in which different cell wall components are deposited has been determined largely by immuno-electron microscopy. The first components to arrive are pectins, hemicelluloses, and arabinogalactan proteins carried by the secretory vesicles that fuse to form the cell plate. The next component to be added is callose, which is polymerized directly at the cell plate by callose synthases. As the cell plate continues to mature and fuses with the parental plasma membrane, the callose is slowly replaced with cellulose, the primary component of a mature cell wall . The middle lamella (a glue-like layer containing pectin) develops from the cell plate, serving to bind the cell walls of adjoining cells together.

Forces in Cytokinesis

Animal Cells

Cytokinetic furrow ingression is powered by Type II Myosin ATPase. Since Myosins are recruited to the medial region, the contractile forces acting on the cortex resemble a 'purse string' constriction pulling inwards. This leads to the inward constriction. The plasma membrane by virtue of its close association with the cortex via crosslinker proteins

Etymology and Pronunciation

The word *cytokinesis* uses combining forms of *cyto-* + *kine-* + *-sis*, New Latin from Classical Latin and Ancient Greek, reflecting "cell" and *kinesis* ("motion, movement").

DNA Replication

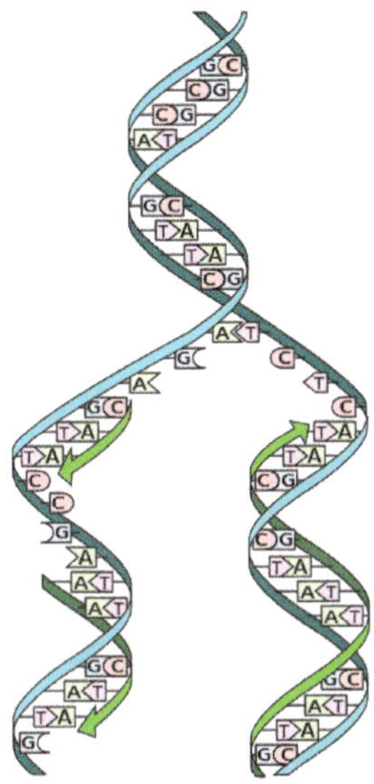

DNA replication. The double helix is unwound and each strand acts as a template for the next strand. Bases are matched to synthesize the new partner strands.

In molecular biology, DNA replication is the biological process of producing two identical replicas of DNA from one original DNA molecule. This process occurs in all living organisms and is the basis for biological inheritance. DNA is made up of a double helix of two complementary strands. During replication, these strands are separated. Each strand of the original DNA molecule then serves as a template for the production of its counterpart, a process referred to as semiconservative replication. Cellular proofreading and error-checking mechanisms ensure near perfect fidelity for DNA replication.

In a cell, DNA replication begins at specific locations, or origins of replication, in the genome. Unwinding of DNA at the origin and synthesis of new strands results in replication forks growing bi-directionally from the origin. A number of proteins are associated with the replication fork to help in the initiation and continuation of DNA synthesis. Most prominently, DNA polymerase synthesizes the new strand by adding nucleotides that complement the template strand. DNA replication occurs during the S-stage of interphase.

DNA replication can also be performed *in vitro* (artificially, outside a cell). DNA polymerases isolated from cells and artificial DNA primers can be used to initiate DNA synthesis at known sequences in a template DNA molecule. The polymerase chain reaction (PCR), a common laboratory technique, cyclically applies such artificial synthesis to amplify a specific target DNA fragment from a pool of DNA.

DNA Structures

DNA usually exists as a double-stranded structure, with both strands coiled together to form the characteristic double-helix. Each single strand of DNA is a chain of four types of nucleotides. Nucleotides in DNA contain a deoxyribose sugar, a phosphate, and a nucleobase. The four types of nucleotide correspond to the four nucleobases adenine, cytosine, guanine, and thymine, commonly abbreviated as A,C, G and T. Adenine and guanine are purine bases, while cytosine and thymine are pyrimidines. These nucleotides form phosphodiester bonds, creating the phosphate-deoxyribose backbone of the DNA double helix with the nuclei bases pointing inward (i.e., toward the opposing strand). Nucleotides (bases) are matched between strands through hydrogen bonds to form base pairs. Adenine pairs with thymine (two hydrogen bonds), and guanine pairs with cytosine (stronger: three hydrogen bonds).

DNA strands have a directionality, and the different ends of a single strand are called the "3' (three-prime) end" and the "5' (five-prime) end". By convention, if the base sequence of a single strand of DNA is given, the left end of the sequence is the 5' end, while the right end of the sequence is the 3' end. The strands of the double helix are anti-parallel with one being 5' to 3', and the opposite strand 3' to 5'. These terms refer to the carbon atom in deoxyribose to which the next phosphate in the chain attaches. Directionality has consequences in DNA synthesis, because DNA polymerase can synthesize DNA in only one direction by adding nucleotides to the 3' end of a DNA strand.

The pairing of complementary bases in DNA (through hydrogen bonding) means that the information contained within each strand is redundant. Phosphodiester bonds are stronger than hydrogen bonds. This allows the strands to be separated from one another. The nucleotides on a single strand can therefore be used to reconstruct nucleotides on a newly synthesized partner strand.

DNA Polymerase

DNA polymerases are a family of enzymes that carry out all forms of DNA replication. DNA polymerases in general cannot initiate synthesis of new strands, but can only extend an existing DNA or RNA strand paired with a template strand. To begin synthesis, a short fragment of RNA, called a primer, must be created and paired with the template DNA strand.

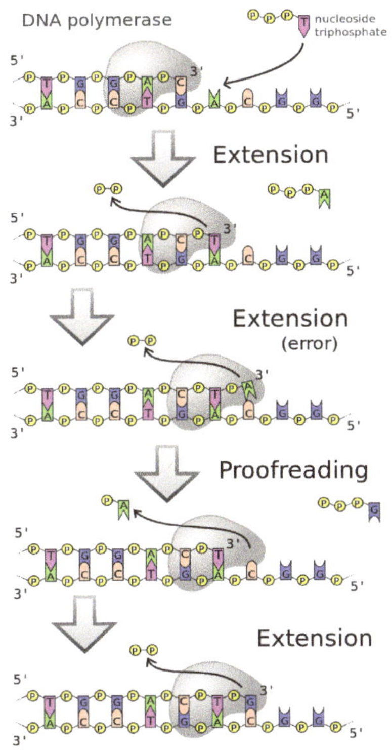

DNA polymerases adds nucleotides to the 3' end of a strand of DNA. If a mismatch is accidentally incorporated, the polymerase is inhibited from further extension. Proofreading removes the mismatched nucleotide and extension continues.

DNA polymerase adds a new strand of DNA by extending the 3' end of an existing nucleotide chain, adding new nucleotides matched to the template strand one at a time via the creation of phosphodiester bonds. The energy for this process of DNA polymerization comes from hydrolysis of the high-energy phosphate (phosphoanhydride) bonds between the three phosphates attached to each unincorporated base. (Free bases with their attached phosphate groups are called nucleotides; in particular, bases with three attached phosphate groups are called nucleoside triphosphates.) When a nucleotide is being added to a growing DNA strand, the formation of a phosphodiester bond between the proximal phosphate of the nucleotide to the growing chain is accompanied by hydrolysis of a high-energy phosphate bond with release of the two distal phosphates as a pyrophosphate. Enzymatic hydrolysis of the resulting pyrophosphate into inorganic phosphate consumes a second high-energy phosphate bond and renders the reaction effectively irreversible.

In general, DNA polymerases are highly accurate, with an intrinsic error rate of less than one mistake for every 10^7 nucleotides added. In addition, some DNA polymerases also have proofreading ability; they can remove nucleotides from the end of a growing strand in order to correct mismatched bases. Finally, post-replication mismatch repair mechanisms monitor the DNA for errors, being capable of distinguishing mismatches in the newly synthesized DNA strand from the original strand sequence. Together, these three discrimination steps enable replication fidelity of less than one mistake for every 10^9 nucleotides added.

The rate of DNA replication in a living cell was first measured as the rate of phage T4 DNA elongation in phage-infected E. coli. During the period of exponential DNA increase at 37 °C, the rate was 749 nucleotides per second. The mutation rate per base pair per replication during phage T4 DNA synthesis is 1.7 per 10^8.

Replication Process

DNA replication, like all biological polymerization processes, proceeds in three enzymatically catalyzed and coordinated steps: initiation, elongation and termination.

Initiation

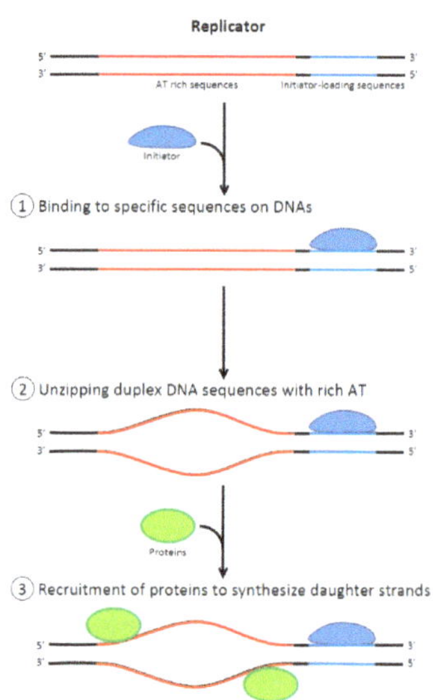

Role of initiators for initiation of DNA replication.

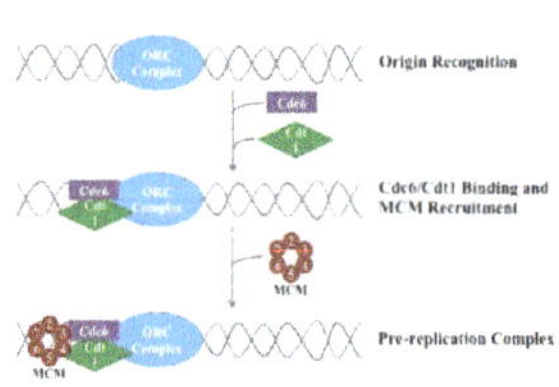

Formation of pre-replication complex.

For a cell to divide, it must first replicate its DNA. This process is initiated at particular points in the DNA, known as "origins", which are targeted by initiator proteins. In *E. coli* this protein is DnaA; in yeast, this is the origin recognition complex. Sequences used by initiator proteins tend to be "AT-rich" (rich in adenine and thymine bases), because A-T base pairs have two hydrogen bonds (rather than the three formed in a C-G pair) which are easier to unzip. Once the origin has been located, these initiators recruit other proteins and form the pre-replication complex, which unzips the double-stranded DNA.

Elongation

DNA polymerase has 5'-3' activity. All known DNA replication systems require a free 3' hydroxyl group before synthesis can be initiated (Important note: the DNA template is read in 3' to 5' direction whereas a new strand is synthesized in the 5' to 3' direction—this is often confused). Four distinct mechanisms for initiation of synthesis are recognized:

All cellular life forms and many DNA viruses, phages and plasmids use a primase to synthesize a short RNA primer with a free 3' OH group which is subsequently elongated by a DNA polymerase.

The retroelements (including retroviruses) employ a transfer RNA that primes DNA replication by providing a free 3′ OH that is used for elongation by the reverse transcriptase.

In the adenoviruses and the φ29 family of bacteriophages, the 3' OH group is provided by the side chain of an amino acid of the genome attached protein (the terminal protein) to which nucleotides are added by the DNA polymerase to form a new strand.

In the single stranded DNA viruses — a group that includes the circoviruses, the geminiviruses, the parvoviruses and others — and also the many phages and plasmids that use the rolling circle replication (RCR) mechanism, the RCR endonuclease creates a nick in the genome strand (single stranded viruses) or one of the DNA strands (plasmids). The 5′ end of the nicked strand is transferred to a tyrosine residue on the nuclease and the free 3′ OH group is then used by the DNA polymerase to synthesize the new strand.

The first is the best known of these mechanisms and is used by the cellular organisms. In this mechanism, once the two strands are separated, primase adds RNA primers to the template strands. The leading strand receives one RNA primer while the lagging strand receives several. The leading strand is continuously extended from the primer by a DNA polymerase with high processivity, while the lagging strand is extended discontinuously from each primer forming Okazaki fragments. RNase removes the primer RNA fragments, and a low processivity DNA polymerase distinct from the replicative polymerase enters to fill the gaps. When this is complete, a single nick on the leading strand and several nicks on the lagging strand can be found. Ligase works to fill these nicks in, thus completing the newly replicated DNA molecule.

The primase used in this process differs significantly between bacteria and archaea/eukaryotes. Bacteria use a primase belonging to the DnaG protein superfamily which contains a catalytic domain of the TOPRIM fold type. The TOPRIM fold contains an α/β core with four conserved strands in a Rossmann-like topology. This structure is also found in the catalytic domains of topoisomerase Ia, topoisomerase II, the OLD-family nucleases and DNA repair proteins related to the RecR protein.

The primase used by archaea and eukaryotes, in contrast, contains a highly derived version of the RNA recognition motif (RRM). This primase is structurally similar to many viral RNA-dependent RNA polymerases, reverse transcriptases, cyclic nucleotide generating cyclases and DNA polymerases of the A/B/Y families that are involved in DNA replication and repair. In eukaryotic replication, the primase forms a complex with Pol α.

Multiple DNA polymerases take on different roles in the DNA replication process. In *E. coli*, DNA Pol III is the polymerase enzyme primarily responsible for DNA replication. It assembles into a replication complex at the replication fork that exhibits extremely high processivity, remaining intact for the entire replication cycle. In contrast, DNA Pol I is the enzyme responsible for replacing RNA primers with DNA. DNA Pol I has a 5' to 3' exonuclease activity in addition to its polymerase activity, and uses its exonuclease activity to degrade the RNA primers ahead of it as it extends the DNA strand behind it, in a process called nick translation. Pol I is much less processive than Pol III because its primary function in DNA replication is to create many short DNA regions rather than a few very long regions.

In eukaryotes, the low-processivity enzyme, Pol α, helps to initiate replication. The high-processivity extension enzymes are Pol δ and Pol ε.

As DNA synthesis continues, the original DNA strands continue to unwind on each side of the bubble, forming a replication fork with two prongs. In bacteria, which have a single origin of replication on their circular chromosome, this process creates a "theta structure" (resembling the Greek letter theta: θ). In contrast, eukaryotes have longer linear chromosomes and initiate replication at multiple origins within these.>

Replication Fork

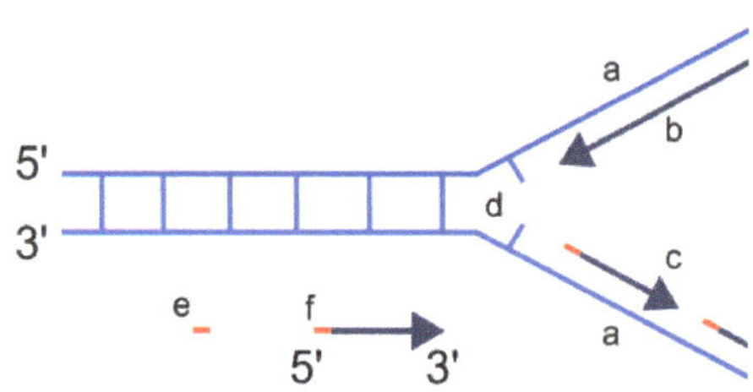

Scheme of the replication fork.

a: template, b: leading strand, c: lagging strand, d: replication fork, e: primer, f: Okazaki fragments

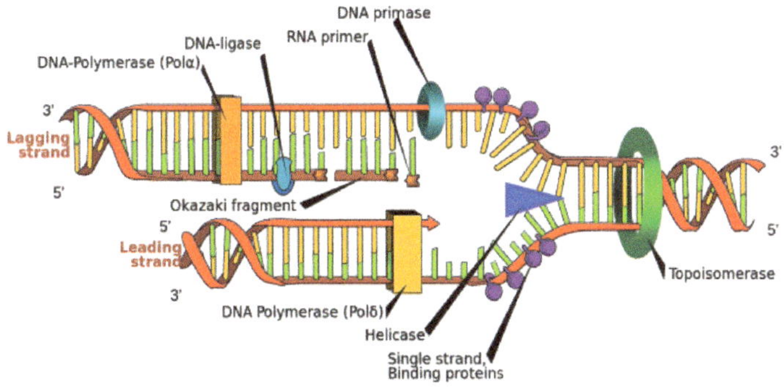

Many enzymes are involved in the DNA replication fork.

The replication fork is a structure that forms within the nucleus during DNA replication. It is created by helicases, which break the hydrogen bonds holding the two DNA strands together. The resulting structure has two branching "prongs", each one made up of a single strand of DNA. These two strands serve as the template for the leading and lagging strands, which will be created as DNA polymerase matches complementary nucleotides to the templates; the templates may be properly referred to as the leading strand template and the lagging strand template.

DNA is always synthesized in the 5' to 3' direction. Since the leading and lagging strand templates are oriented in opposite directions at the replication fork, a major issue is how to achieve synthesis of nascent (new) lagging strand DNA, whose direction of synthesis is opposite to the direction of the growing replication fork.

Leading Strand

The leading strand is the strand of nascent DNA which is being synthesized in the same direction as the growing replication fork. A polymerase "reads" the leading strand *template* and adds complementary nucleotides to the nascent leading strand on a continuous basis.

The polymerase involved in leading strand synthesis is DNA polymerase III (DNA Pol III) in prokaryotes. In eukaryotes, leading strand synthesis is thought to be conducted by Pol ε, however this view has been recently challenged, suggesting a role for Pol δ.

Lagging Strand

The lagging strand is the strand of nascent DNA whose direction of synthesis is opposite to the direction of the growing replication fork. Because of its orientation, replication of the lagging strand is more complicated as compared to that of the leading strand.

The lagging strand is synthesized in short, separated segments. On the lagging strand *template*, a primase "reads" the template DNA and initiates synthesis of a short complementary RNA primer. A DNA polymerase extends the primed segments, forming

Okazaki fragments. The RNA primers are then removed and replaced with DNA, and the fragments of DNA are joined together by DNA ligase.

DNA polymerase III (in prokaryotes) or Pol δ (in eukaryotes) is responsible for extension of the primers added during replication of the lagging strand. Primer removal is performed by DNA polymerase I (in prokaryotes) and Pol δ (in eukaryotes). Eukaryotic primase is intrinsic to Pol α. In eukaryotes, pol ε helps with repair during DNA replication.

Dynamics at the Replication Fork

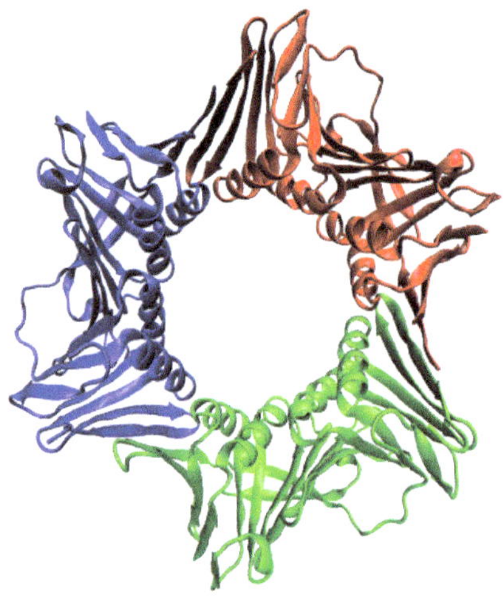

The assembled human DNA clamp, a trimer of the protein PCNA.

As helicase unwinds DNA at the replication fork, the DNA ahead is forced to rotate. This process results in a build-up of twists in the DNA ahead. This build-up forms a torsional resistance that would eventually halt the progress of the replication fork. Topoisomerases are enzymes that temporarily break the strands of DNA, relieving the tension caused by unwinding the two strands of the DNA helix; topoisomerases (including DNA gyrase) achieve this by adding negative supercoils to the DNA helix.

Bare single-stranded DNA tends to fold back on itself forming secondary structures; these structures can interfere with the movement of DNA polymerase. To prevent this, single-strand binding proteins bind to the DNA until a second strand is synthesized, preventing secondary structure formation.

Clamp proteins form a sliding clamp around DNA, helping the DNA polymerase maintain contact with its template, thereby assisting with processivity. The inner face of the clamp enables DNA to be threaded through it. Once the polymerase reaches the end of the template or detects double-stranded DNA, the sliding clamp undergoes a conformational change that releases the DNA polymerase. Clamp-loading proteins are used to initially load the clamp, recognizing the junction between template and RNA primers.

DNA Replication Proteins

At the replication fork, many replication enzymes assemble on the DNA into a complex molecular machine called the replisome. The following is a list of major DNA replication enzymes that participate in the replisome:

Enzyme	Function in DNA replication
DNA Helicase	Also known as helix destabilizing enzyme. Unwinds the DNA double helix at the Replication Fork.
DNA Polymerase	Builds a new duplex DNA strand by adding nucleotides in the 5' to 3' direction. Also performs proof-reading and error correction. There exist many different types of DNA Polymerase, each of which perform different functions in different types of cells.
DNA clamp	A protein which prevents elongating DNA polymerases from dissociating from the DNA parent strand.
Single-Strand Binding (SSB) Proteins	Bind to ssDNA and prevent the DNA double helix from re-annealing after DNA helicase unwinds it, thus maintaining the strand separation, and facilitating the synthesis of the nascent strand.
Topoisomerase	Relaxes the DNA from its super-coiled nature.
DNA Gyrase	Relieves strain of unwinding by DNA helicase; this is a specific type of topoisomerase
DNA Ligase	Re-anneals the semi-conservative strands and joins Okazaki Fragments of the lagging strand.
Primase	Provides a starting point of RNA (or DNA) for DNA polymerase to begin synthesis of the new DNA strand.
Telomerase	Lengthens telomeric DNA by adding repetitive nucleotide sequences to the ends of eukaryotic chromosomes. This allows germ cells and stem cells to avoid the Hayflick limit on cell division.

Replication Machinery

Replication machineries consist of factors involved in DNA replication and appearing on template ssDNAs. Replication machineries include primosotors are replication enzymes; DNA polymerase, DNA helicases, DNA clamps and DNA topoisomerases, and replication proteins; e.g. single-stranded DNA binding proteins (SSB). In the replication machineries these components coordinate. In most of the bacteria, all of the factors involved in DNA replication are located on replication forks and the complexes stay on the forks during DNA replication. This replication machineries are called replisomes or DNA replicase systems, these terms mean originally a generic term for proteins located on replication forks. In eukaryotic and some bacterial cells the replisomes are not formed.

Since replication machineries do not move relatively to template DNAs such as factories, they are called replication factory. In an alternative figure, DNA factories are similar to projectors and DNAs are like as cinematic films passing constantly into the projectors. In the replication factory model, after both DNA helicases for leading stands and lagging strands are loaded on the template DNAs, the helicases run along the DNAs into each other. The helicases remain associated for the remainder of replication process. Peter Meister et al. observed directly replication sites in budding yeast by monitoring green fluorescent protein(GFP)-tagged DNA polymerases α. They detected DNA replication of pairs of the tagged loci spaced apart symmetrically from a replication origin and found that the distance between the pairs decreased markedly by time. This finding suggests that the mechanism of DNA replication goes with DNA factories. That is, couples of replication factories are loaded on replication origins and the factories associated with each other. Also, template DNAs move into the factories, which bring extrusion of the template ssDNAs and nascent DNAs. Meister's finding is the first direct evidence of replication factory model. Subsequent research has shown that DNA helicases form dimers in many eukaryotic cells and bacterial replication machineries stay in single intranuclear location during DNA synthesis.

The replication factories perform disentanglement of sister chromatids. The disentanglement is essential for distributing the chromatids into daughter cells after DNA replication. Because sister chromatids after DNA replication hold each other by Cohesin rings, there is the only chance for the disentanglement in DNA replication. Fixing of replication machineries as replication factories can improve the success rate of DNA replication. If replication forks move freely in chromosomes, catenation of nuclei is aggravated and impedes mitotic segregation.

Termination

Eukaryotes initiate DNA replication at multiple points in the chromosome, so replication forks meet and terminate at many points in the chromosome; these are not known to be regulated in any particular way. Because eukaryotes have linear chromosomes, DNA replication is unable to reach the very end of the chromosomes, but ends at the telomere region of repetitive DNA close to the ends. This shortens the telomere of the daughter DNA strand. Shortening of the telomeres is a normal process in somatic cells. As a result, cells can only divide a certain number of times before the DNA loss prevents further division. (This is known as the Hayflick limit.) Within the germ cell line, which passes DNA to the next generation, telomerase extends the repetitive sequences of the telomere region to prevent degradation. Telomerase can become mistakenly active in somatic cells, sometimes leading to cancer formation. Increased telomerase activity is one of the hallmarks of cancer.

Termination requires that the progress of the DNA replication fork must stop or be blocked. Termination at a specific locus, when it occurs, involves the interaction between two components: (1) a termination site sequence in the DNA, and (2) a protein which binds to this sequence to physically stop DNA replication. In various bacterial species, this is named the DNA replication terminus site-binding protein, or Ter protein.

Because bacteria have circular chromosomes, termination of replication occurs when the two replication forks meet each other on the opposite end of the parental chromosome. *E. coli* regulates this process through the use of termination sequences that, when bound by the Tus protein, enable only one direction of replication fork to pass through. As a result, the replication forks are constrained to always meet within the termination region of the chromosome.

Regulation

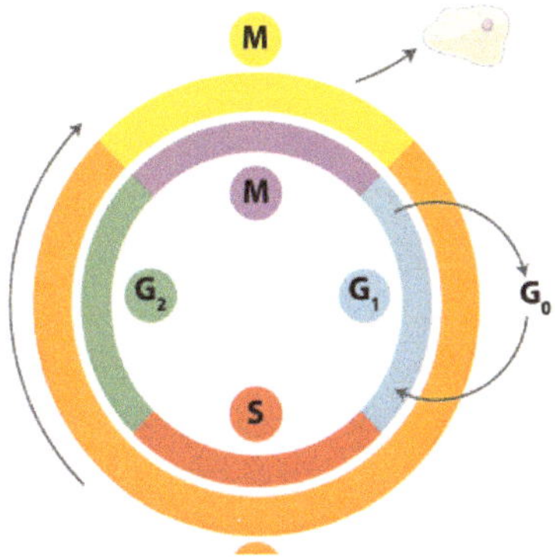

The cell cycle of eukaryotic cells.

Eukaryotes

Within eukaryotes, DNA replication is controlled within the context of the cell cycle. As the cell grows and divides, it progresses through stages in the cell cycle; DNA replication takes place during the S phase (synthesis phase). The progress of the eukaryotic cell through the cycle is controlled by cell cycle checkpoints. Progression through checkpoints is controlled through complex interactions between various proteins, including cyclins and cyclin-dependent kinases. Unlike bacteria, eukaryotic DNA replicates in the confines of the nucleus.

The G1/S checkpoint (or restriction checkpoint) regulates whether eukaryotic cells enter the process of DNA replication and subsequent division. Cells that do not proceed through this checkpoint remain in the Go stage and do not replicate their DNA.

Replication of chloroplast and mitochondrial genomes occurs independently of the cell cycle, through the process of D-loop replication.

Replication Focus

In vertebrate cells, replication sites concentrate into positions called replication foci. Replication sites can be detected by immunostaining daughter strands and replication enzymes and monitoring GFP-tagged replication factors. By these methods it is found that replication foci of varying size and positions appear in S phase of cell division and their number per nucleus is far smaller than the number of genomic replication forks.

P. Heun et al.(2001) tracked GFP-tagged replication foci in budding yeast cells and revealed that replication origins move constantly in G1 and S phase and the dynamics decreased significantly in S phase. Traditionally, replication sites were fixed on spatial structure of chromosomes by nuclear matrix or lamins. The Heun's results denied the traditional concepts, budding yeasts don't have lamins, and support that replication origins self-assemble and form replication foci.

By firing of replication origins, controlled spatially and temporally, the formation of replication foci is regulated. D. A. Jackson et al.(1998) revealed that neighboring origins fire simultaneously in mammalian cells. Spatial juxtaposition of replication sites brings clustering of replication forks. The clustering do rescue of stalled replication forks and favors normal progress of replication forks. Progress of replication forks is inhibited by many factors; collision with proteins or with complexes binding strongly on DNA, deficiency of dNTPs, nicks on template DNAs and so on. If replication forks stall and remaining sequences from the stalled forks are not replicated, daughter strands have nick obtained un-replicated sites. The un-replicated sites on one parent's strand hold the other strand together but not daughter strands. Therefore, resulting sister chromatids cannot separate together and cannot divide into 2 daughter cells. When neighboring origins fire and a fork from one origin is stalled, fork from other origin access on an opposite direction of the stalled fork and duplicate the un-replicated sites. As other mechanism of the rescue there is application of dormant replication origins that excess origins don't fire in normal DNA replication.

Bacteria

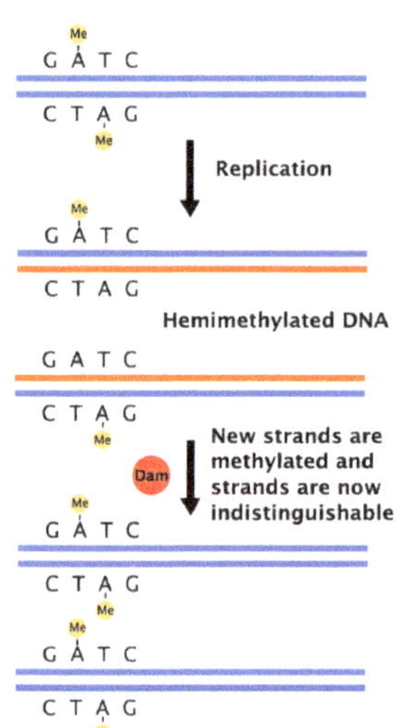

Dam methylates adenine of GATC sites after replication.

Most bacteria do not go through a well-defined cell cycle but instead continuously copy their DNA; during rapid growth, this can result in the concurrent occurrence of multiple rounds of replication. In *E. coli*, the best-characterized bacteria, DNA replication is regulated through several mechanisms, including: the hemimethylation and sequestering of the origin sequence, the ratio of adenosine triphosphate (ATP) to adenosine diphosphate (ADP), and the levels of protein DnaA. All these control the binding of initiator proteins to the origin sequences.

Because *E. coli* methylates GATC DNA sequences, DNA synthesis results in hemimethylated sequences. This hemimethylated DNA is recognized by the protein SeqA, which binds and sequesters the origin sequence; in addition, DnaA (required for initiation of replication) binds less well to hemimethylated DNA. As a result, newly replicated origins are prevented from immediately initiating another round of DNA replication.

ATP builds up when the cell is in a rich medium, triggering DNA replication once the cell has reached a specific size. ATP competes with ADP to bind to DnaA, and the DnaA-ATP complex is able to initiate replication. A certain number of DnaA proteins are also required for DNA replication – each time the origin is copied, the number of binding sites for DnaA doubles, requiring the synthesis of more DnaA to enable another initiation of replication.

Polymerase Chain Reaction

Researchers commonly replicate DNA *in vitro* using the polymerase chain reaction (PCR). PCR uses a pair of primers to span a target region in template DNA, and then polymerizes partner strands in each direction from these primers using a thermostable DNA polymerase. Repeating this process through multiple cycles amplifies the targeted DNA region. At the start of each cycle, the mixture of template and primers is heated, separating the newly synthesized molecule and template. Then, as the mixture cools, both of these become templates for annealing of new primers, and the polymerase extends from these. As a result, the number of copies of the target region doubles each round, increasing exponentially.

Homologous Recombination

Homologous recombination is a type of genetic recombination in which nucleotide sequences are exchanged between two similar or identical molecules of DNA. It is most widely used by cells to accurately repair harmful breaks that occur on both strands of DNA, known as double-strand breaks. Homologous recombination also produces new combinations of DNA sequences during meiosis, the process by which eukaryotes make gamete cells, like sperm and egg cells in animals. These new combinations of DNA represent genetic variation in offspring, which in turn enables populations to adapt during the course of evolution. Homologous recombination is also used in horizontal gene transfer to exchange genetic material between different strains and species of bacteria and viruses.

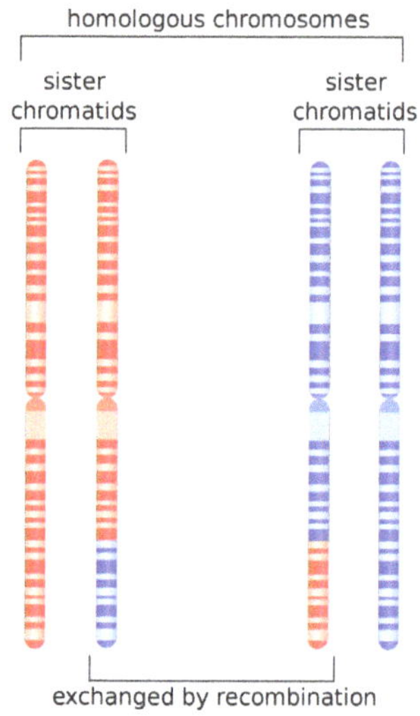

Figure. During meiosis, homologous recombination can produce new combinations of genes as shown here between similar but not identical copies of human chromosome 1.

Although homologous recombination varies widely among different organisms and cell types, most forms involve the same basic steps. After a double-strand break occurs, sections of DNA around the 5' ends of the break are cut away in a process called *resection*. In the *strand invasion* step that follows, an overhanging 3' end of the broken DNA molecule then "invades" a similar or identical DNA molecule that is not broken. After strand invasion, the further sequence of events may follow either of two main pathways discussed below; the DSBR (double-strand break repair) pathway or the SDSA (synthesis-dependent strand annealing) pathway. Homologous recombination that occurs during DNA repair tends to result in non-crossover products, in effect restoring the damaged DNA molecule as it existed before the double-strand break.

Homologous recombination is conserved across all three domains of life as well as viruses, suggesting that it is a nearly universal biological mechanism. The discovery of genes for homologous recombination in protists—a diverse group of eukaryotic microorganisms—has been interpreted as evidence that meiosis emerged early in the evolution of eukaryotes. Since their dysfunction has been strongly associated with increased susceptibility to several types of cancer, the proteins that facilitate homologous recombination are topics of active research. Homologous recombination is also used in gene targeting, a technique for introducing genetic changes into target organisms. For their development of this technique, Mario Capecchi, Martin Evans and Oliver Smithies were awarded the 2007 Nobel Prize for Physiology or Medicine; Capecchi and Smithies independently discovered applications to mouse embryonic stem cells, however the highly conserved mechanisms underlying the DSB repair model, including uniform homologous integration of transformed DNA (gene therapy), were first shown in plasmid experiments by Orr-Weaver, Szostack and Rothstein. Researching the plasmid-induced DSB, using γ-irradiation in the 1970's-1980's, led to later experiments using endonucleases (e.g. I-SceI) to cut chromosomes for genetic engineering of mammalian cells, where nonhomologous recombination is more frequent than in yeast.

History and Discovery

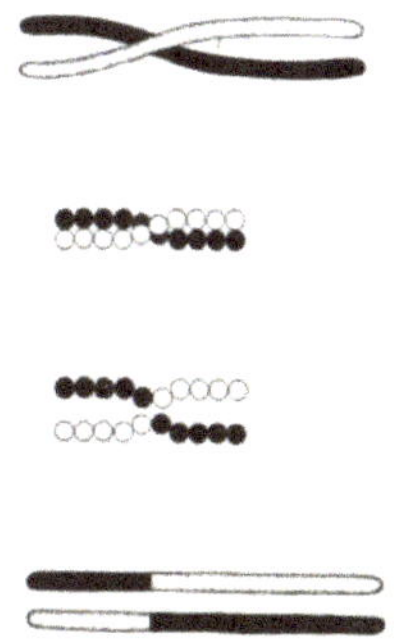

Fig. 64. Scheme to illustrate a method of crossing over of the chromosomes.

Figure. An early illustration of crossing over from Thomas Hunt Morgan

In the early 1900s, William Bateson and Reginald Punnett found an exception to one of the principles of inheritance originally described by Gregor Mendel in the 1860s. In contrast to Mendel's notion that traits are independently assorted when passed from parent to child—for example that a cat's hair color and its tail length are inherited independent of each other—Bateson and Punnett showed that certain genes associated with physical traits can be inherited together, or genetically linked. In 1911, after observing that linked traits could on occasion be inherited separately, Thomas Hunt Morgan suggested that "crossovers" can occur between linked genes, where one of the linked genes physically crosses over to a different chromosome. Two decades later, Barbara McClintock and Harriet Creighton demonstrated that chromosomal crossover occurs during meiosis, the process of cell division by which sperm and egg cells are made. Within the same year as McClintock's discovery, Curt Stern showed that crossing over—later called "recombination"—could also occur in somatic cells like white blood cells and skin cells that divide through mitosis.

In 1947, the microbiologist Joshua Lederberg showed that bacteria—which had been assumed to reproduce only asexually through binary fission—are capable of genetic recombination, which is more similar to sexual reproduction. This work established *E. coli* as a model organism in genetics, and helped Lederberg win the 1958 Nobel Prize in Physiology or Medicine. Building on studies in fungi, in 1964 Robin Holliday proposed a model for recombination in meiosis which introduced key details of how the process can work, including the exchange of material between chromosomes through Holliday junctions. In 1983, Jack Szostak and colleagues presented a model now known as the DSBR pathway, which accounted for observations not explained by the Holliday model. During the next decade, experiments in *Drosophila*, budding yeast and mammalian cells led to the emergence of other models of homologous recombination, called SDSA pathways, which do not always rely on Holliday junctions.

In Eukaryotes

Homologous recombination (HR) is essential to cell division in eukaryotes like plants, animals, fungi and protists. In cells that divide through mitosis, homologous recombination repairs double-strand breaks in DNA caused by ionizing radiation or DNA-damaging chemicals. Left unrepaired, these double-strand breaks can cause large-scale rearrangement of chromosomes in somatic cells, which can in turn lead to cancer.

In addition to repairing DNA, homologous recombination also helps produce genetic diversity when cells divide in meiosis to become specialized gamete cells—sperm or egg cells in animals, pollen or ovules in plants, and spores in fungi. It does so by facilitating chromosomal crossover, in which regions of similar but not identical DNA are exchanged between homologous chromosomes. This creates new, possibly beneficial combinations of genes, which can give offspring an evolutionary advantage. Chromosomal crossover often begins when a protein called Spo11 makes a targeted double-strand break in DNA. These sites are non-randomly located on the chromosomes; usually in intergenic promoter regions and preferentially in GC-rich domains These double-strand break sites often occur at recombination hotspots, regions in chromosomes that are about 1,000–2,000 base pairs in length and have high rates of recombination. The absence of a recombination hotspot between two genes on the same chromosome often means that those genes will be inherited by future generations in equal proportion. This represents linkage between the two genes greater than would be expected from genes that independently assort during meiosis.

Timing within the Mitotic Cell Cycle

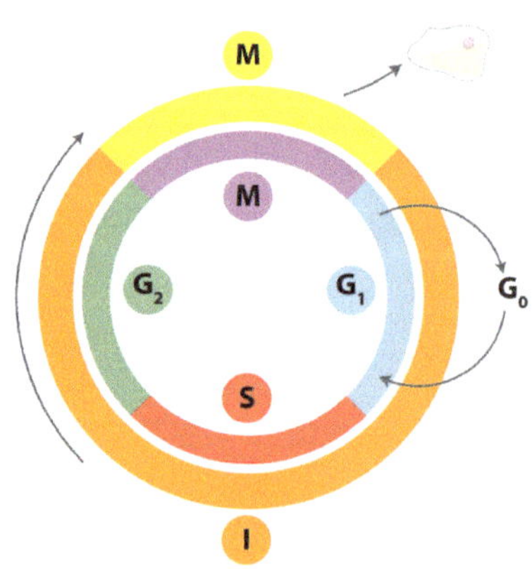

Figure. Homologous recombination repairs DNA before the cell enters mitosis (M phase). It occurs only during and shortly after DNA replication, during the S and G_2 phases of the cell cycle.

Double-strand breaks can be repaired through homologous recombination or through non-homologous end joining (NHEJ). NHEJ is a DNA repair mechanism which, unlike homologous recombination, does not require a long homologous sequence to guide repair. Whether homologous recombination or NHEJ is used to repair double-strand breaks is largely determined by the phase of cell cycle. Homologous recombination repairs DNA before the cell enters mitosis (M phase). It occurs during and shortly after DNA replication, in the S and G_2 phases of the cell cycle, when sister chromatids are more easily available. Compared to homologous chromosomes, which are similar to another chromosome but often have different alleles, sister chromatids are an ideal template for homologous recombination because they are an identical copy of a given chromosome. In contrast to homologous recombination, NHEJ is predominant in the G_1 phase of the cell cycle, when the cell is growing but not yet ready to divide. It occurs less frequently after the G_1 phase, but maintains at least some activity throughout the cell cycle. The mechanisms that regulate homologous recombination and NHEJ throughout the cell cycle vary widely between species.

Cyclin-dependent kinases (CDKs), which modify the activity of other proteins by adding phosphate groups to (that is, phosphorylating) them, are important regulators of homologous recombination in eukaryotes. When DNA replication begins in budding yeast, the cyclin-dependent kinase Cdc28 begins homologous recombination by phosphorylating the Sae2 protein. After being so activated by the addition of a phosphate, Sae2 uses its endonuclease activity to make a clean cut near a double-strand break in DNA. This allows a three-part protein known as the MRX complex to bind to DNA, and begins a series of protein-driven reactions that exchange material between two DNA molecules.

Models

Two primary models for how homologous recombination repairs double-strand breaks in DNA are the double-strand break repair (DSBR) pathway (sometimes called the *double Holliday junction model*) and the synthesis-dependent strand annealing (SDSA) pathway. The two pathways are similar in their first several steps. After a double-strand break occurs, the MRX complex (MRN complex in humans) binds to DNA on either side of the break. Next a resection, in which DNA around the 5' ends of the break is cut back, is carried out in two distinct steps. In the first step of resection, the MRX complex recruits the Sae2 protein. The two proteins then trim back the 5' ends on either side of the break to create short 3' overhangs of single-strand DNA. In the second step, 5'→3' resection is continued by the Sgs1 helicase and the Exo1 and Dna2 nucleases. As a helicase, Sgs1 "unzips" the double-strand DNA, while Exo1 and Dna2's nuclease activity allows them to cut the single-stranded DNA produced by Sgs1.

The RPA protein, which has high affinity for single-stranded DNA, then binds the 3' overhangs. With the help of several other proteins that mediate the process, the Rad51 protein (and Dmc1, in meiosis) then forms a filament of nucleic acid and protein on the single strand of DNA coated with RPA. This nucleoprotein filament then begins searching for DNA sequences similar to that of the 3' overhang. After finding such a sequence, the single-stranded nucleoprotein filament moves into (invades) the similar or identical recipient DNA duplex in a process called *strand invasion*. In cells that divide through mitosis, the recipient DNA duplex is generally a sister chromatid, which is identical to the damaged DNA molecule and provides a template for repair. In meiosis, however, the recipient DNA tends to be from a similar but not necessarily identical homologous chromosome. A displacement loop (D-loop) is formed during strand invasion between the invading 3' overhang strand and the homologous chromosome. After strand invasion, a DNA polymerase extends the end of the invading 3' strand by synthesizing new DNA. This changes the D-loop to a cross-shaped structure known as a Holliday junction. Following this, more DNA synthesis occurs on the invading strand (i.e., one of the original 3' overhangs), effectively restoring the strand on the homologous chromosome that was displaced during strand invasion.

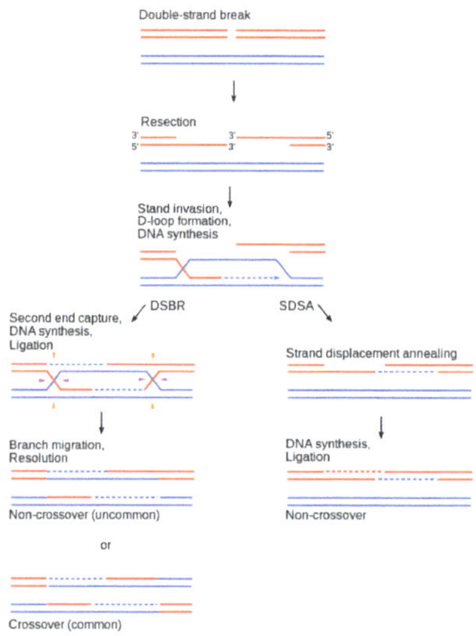

Figure. The DSBR and SDSA pathways follow the same initial steps, but diverge thereafter. The DSBR pathway most often results in chromosomal crossover (bottom left), while SDSA always ends with non-crossover products (bottom right).

DSBR Pathway

After the stages of resection, strand invasion and DNA synthesis, the DSBR and SDSA pathways become distinct. The DSBR pathway is unique in that the second 3' overhang (which was not involved in strand invasion) also forms a Holliday junction with the homologous chromosome. The double Holliday junctions are then converted into recombination products by nicking endonucleases, a type of restriction endonuclease which cuts only one DNA strand. The DSBR pathway commonly results in crossover, though it can sometimes result in non-crossover products; the ability of a broken DNA molecule to collect sequences from separated donor loci was shown in mitotic budding yeast using plasmids or endonuclease induction of chromosomal events. Because of this tendency for chromosomal crossover, the DSBR pathway is a likely model of how crossover homologous recombination occurs during meiosis.

Whether recombination in the DSBR pathway results in chromosomal crossover is determined by how the double Holliday junction is cut, or "resolved". Chromosomal crossover will occur if one Holliday junction is cut on the crossing strand and the other Holliday junction is cut on the non-crossing strand (in Figure, along the horizontal purple arrowheads at one Holliday junction and along the vertical orange arrowheads at the other). Alternatively, if the two Holliday junctions are cut on the crossing strands (along the horizontal purple arrowheads at both Holliday junctions in Figure), then chromosomes without crossover will be produced.

SDSA Pathway

Homologous recombination via the SDSA pathway occurs in cells that divide through mitosis and meiosis and results in non-crossover products. In this model, the invading 3' strand is extended along the recipient DNA duplex by a DNA polymerase, and is released as the Holliday junction between the donor and recipient DNA molecules slides in a process called *branch migration*. The newly synthesized 3' end of the invading strand is then able to anneal to the other 3' overhang in the damaged chromosome through complementary base pairing. After the strands anneal, a small flap of DNA can sometimes remain. Any such flaps are removed, and the SDSA pathway finishes with the resealing, also known as *ligation*, of any remaining single-stranded gaps.

During mitosis, the major homologous recombination pathway for repairing DNA double-strand breaks appears to be the SDSA pathway (rather than the DSBR pathway). The SDSA pathway produces non-crossover recombinants (Figure). During meiosis non-crossover recombinants also occur frequently and these appear to arise mainly by the SDSA pathway as well. Non-crossover recombination events occurring during meiosis likely reflect instances of repair of DNA double-strand damages or other types of DNA damages.

SSA Pathway

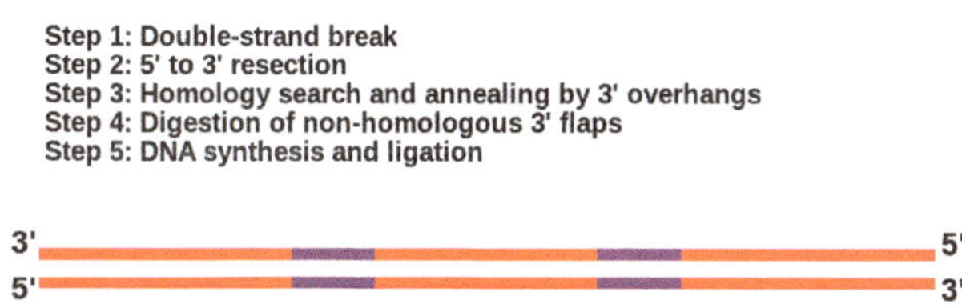

Figure. Recombination via the SSA pathway occurs between two repeat elements (purple) on the same DNA duplex, and results in deletions of genetic material. (Click to view animated diagram in Firefox, Chrome, Safari, or Opera web browsers.)

The single-strand annealing (SSA) pathway of homologous recombination repairs double-strand breaks between two repeat sequences. The SSA pathway is unique in that it does not require a separate similar or identical molecule of DNA, like the DSBR or SDSA pathways of homologous recombination. Instead, the SSA pathway only requires a single DNA duplex, and uses the repeat sequences as the identical sequences that homologous recombination needs for repair. The pathway is relatively simple in concept: after two strands of the same DNA duplex are cut back around the site of the double-strand break, the two resulting 3' overhangs then align and anneal to each other, restoring the DNA as a continuous duplex.

As DNA around the double-strand break is cut back, the single-stranded 3' overhangs being produced are coated with the RPA protein, which prevents the 3' overhangs from sticking to themselves. A protein called Rad52 then binds each of the repeat sequences on either side of the break, and aligns them to enable the two complementary repeat sequences to anneal. After annealing is complete, leftover non-homologous flaps of the 3' overhangs are cut away by a set of nucleases, known as Rad1/Rad10, which are brought to the flaps by the Saw1 and Slx4 proteins. New DNA synthesis fills in any gaps, and ligation restores the DNA duplex as two continuous strands. The DNA sequence between the repeats is always lost, as is one of the two repeats. The SSA pathway is considered mutagenic since it results in such deletions of genetic material.

BIR Pathway

During DNA replication, double-strand breaks can sometimes be encountered at replication forks as DNA helicase unzips the template strand. These defects are repaired in the *break-induced replication* (BIR) pathway of homologous recombination. The precise molecular mechanisms of the BIR pathway remain unclear. Three proposed mechanisms have strand invasion as an initial step, but they differ in how they model the migration of the D-loop and later phases of recombination.

The BIR pathway can also help to maintain the length of telomeres (regions of DNA at the end of eukaryotic chromosomes) in the absence of (or in cooperation with) telomerase. Without working copies of the telomerase enzyme, telomeres typically shorten with each cycle of mitosis, which eventually blocks cell division and leads to senescence. In budding yeast cells where telomerase has been inactivated through mutations, two types of "survivor" cells have been observed to avoid senescence longer than expected by elongating their telomeres through BIR pathways.

Maintaining telomere length is critical for cell immortalization, a key feature of cancer. Most cancers maintain telomeres by upregulating telomerase. However, in several types of human cancer, a BIR-like pathway helps to sustain some tumors by acting as an alternative mechanism of telomere maintenance. This fact has led scientists to investigate whether such recombination-based mechanisms of telomere maintenance could thwart anti-cancer drugs like telomerase inhibitors.

In Bacteria

Homologous recombination is a major DNA repair process in bacteria. It is also important for producing genetic diversity in bacterial populations, although the process differs substantially from meiotic recombination, which repairs DNA damages and brings about diversity in eukaryotic genomes. Homologous recombination has been most studied and is best understood for *Escherichia coli.* Double-strand DNA breaks in bacteria are repaired by the RecBCD pathway of homologous recombination. Breaks that occur on only one of the two DNA strands, known as single-strand gaps, are thought to be repaired by the RecF pathway. Both the RecBCD and RecF pathways include a series of reactions known as *branch migration*, in which single DNA strands are exchanged between two intercrossed molecules of duplex DNA, and *resolution*, in which those two intercrossed molecules of DNA are cut apart and restored to their normal double-stranded state.

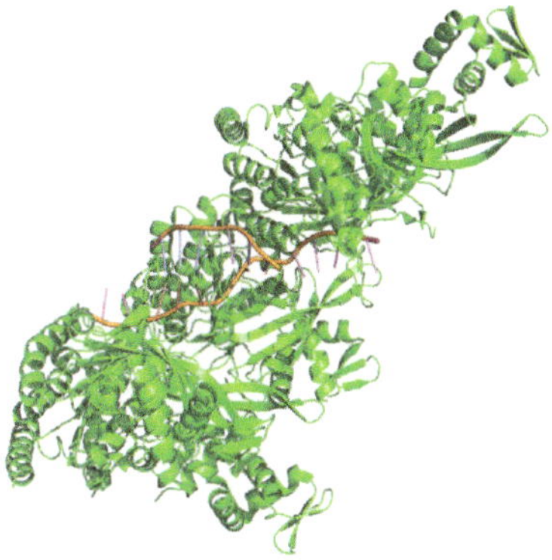

Figure. Crystal structure of a RecA protein filament bound to DNA.
A 3' overhang is visible to the right of center.

RecBCD Pathway

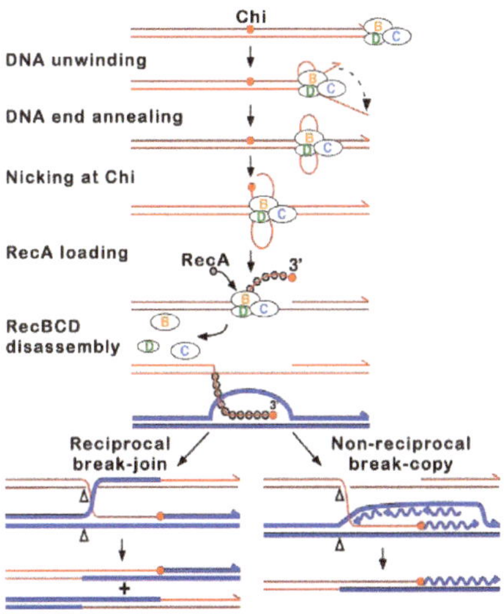

Figure. Molecular model for the RecBCD pathway of recombination. This model is based on reactions of DNA and RecBCD with ATP in excess over Mg2+ ions. Step 1: RecBCD binds to a double-stranded DNA end. Step 2: RecBCD unwinds DNA. RecD is a fast helicase on the 5'-ended strand, and RecB is a slower helicase on the 3'-ended strand (that with an arrowhead). This produces two single-stranded (ss) DNA tails and one ss loop. The loop and tails enlarge as RecBCD moves along the DNA. Step 3: The two tails anneal to produce a second ss DNA loop, and both loops move and grow. Step 4: Upon reaching the Chi hotspot sequence (5' GCTGGTGG 3'; red dot) RecBCD nicks the 3'-ended strand. Further unwinding produces a long 3'-ended ss tail with Chi near its end. Step 5: RecBCD loads RecA protein onto the Chi tail. At some undetermined point, the RecBCD subunits disassemble. Step 6: The RecA-ssDNA complex invades an intact homologous duplex DNA to produce a D-loop, which can be resolved into intact, recombinant DNA in two ways. Step 7: The D-loop is cut and anneals with the gap in the first DNA to produce a Holliday junction. Resolution of the Holliday junction (cutting, swapping of strands, and ligation) at the open arrowheads by some combination of RuvABC and RecG produces two recombinants of reciprocal type. Step 8: The 3' end of the Chi tail primes DNA synthesis, from which a replication fork can be generated. Resolution of the fork at the open arrowheads produces one recombinant (non-reciprocal) DNA, one parental-type DNA, and one DNA fragment.

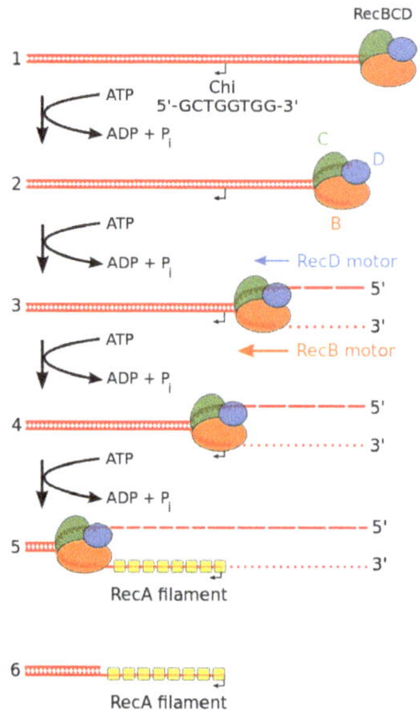

Figure. Beginning of the RecBCD pathway. This model is based on reactions of DNA and RecBCD with Mg^{2+} ions in excess over ATP. Step 1: RecBCD binds to a DNA double strand break. Step 2: RecBCD initiates unwinding of the DNA duplex through ATP-dependent helicase activity. Step 3: RecBCD continues its unwinding and moves down the DNA duplex, cleaving the 3' strand much more frequently than the 5' strand. Step 4: RecBCD encounters a Chi sequence and stops digesting the 3' strand; cleavage of the 5' strand is significantly increased. Step 5: RecBCD loads RecA onto the 3' strand. Step 6: RecBCD unbinds from the DNA duplex, leaving a RecA nucleoprotein filament on the 3' tail.

The RecBCD pathway is the main recombination pathway used in many bacteria to repair double-strand breaks in DNA, and the proteins are found in a broad array of bacteria. These double-strand breaks can be caused by UV light and other radiation, as well as chemical mutagens. Double-strand breaks may also arise by DNA replication through a single-strand nick or gap. Such a situation causes what is known as a collapsed replication fork and is fixed by several pathways of homologous recombination including the RecBCD pathway.

In this pathway, a three-subunit enzyme complex called RecBCD initiates recombination by binding to a blunt or nearly blunt end of a break in double-strand DNA. After RecBCD binds the DNA end, the RecB and RecD subunits begin unzipping the DNA duplex through helicase activity. The RecB subunit also has a nuclease domain, which cuts the single strand of DNA that emerges from the unzipping process. This unzipping continues until RecBCD encounters a specific nucleotide sequence (5'-GCTGGTGG-3') known as a Chi site.

Upon encountering a Chi site, the activity of the RecBCD enzyme changes drastically. DNA unwinding pauses for a few seconds and then resumes at roughly half the initial speed. This is likely because the slower RecB helicase unwinds the DNA after Chi, rather than the faster RecD helicase, which unwinds the DNA before Chi. Recognition of the Chi site also changes the RecBCD enzyme so that it cuts the DNA strand with Chi and begins loading multiple RecA proteins onto the single-stranded DNA with the newly generated 3' end. The resulting RecA-coated nucleoprotein filament then searches out similar sequences of DNA on a homologous chromosome. The search process induces stretching of the DNA duplex, which enhances homology recognition (a mechanism termed conformational proofreading). Upon finding such a sequence, the single-stranded nucleoprotein filament moves into the homologous recipient DNA duplex in a process called *strand invasion*. The invading 3' overhang causes one of the strands of the recipient DNA duplex to be displaced, to form a D-loop. If the D-loop is cut, another swapping of strands forms a cross-shaped structure called a Holliday junction. Resolution of the Holliday junction by some combination of RuvABC or RecG can produce two recombinant DNA molecules with reciprocal genetic types, if the two interacting DNA molecules differ genetically. Alternatively, the invading 3' end near Chi can prime DNA synthesis and form a replication fork. This type of resolution produces only one type of recombinant (non-reciprocal).

RecF Pathway

Bacteria appear to use the RecF pathway of homologous recombination to repair single-strand gaps in DNA. When the RecBCD pathway is inactivated by mutations and additional mutations inactivate the SbcCD and ExoI nucleases, the RecF pathway can also repair DNA double-strand breaks. In the RecF pathway the RecQ helicase unwinds the DNA and the RecJ nuclease degrades the strand with a 5' end, leaving the strand with the 3' end intact. RecA protein binds to this strand and is either aided by the RecF, RecO, and RecR proteins or stabilized by them. The RecA nucleoprotein filament then searches for a homologous DNA and exchanges places with the identical or nearly identical strand in the homologous DNA.

Although the proteins and specific mechanisms involved in their initial phases differ, the two pathways are similar in that they both require single-stranded DNA with a 3' end and the RecA protein for strand invasion. The pathways are also similar in their phases of *branch migration*, in which the Holliday junction slides in one direction, and *resolution*, in which the Holliday junctions are cleaved apart by enzymes. The alternative, non-reciprocal type of resolution may also occur by either pathway.

Branch Migration

Immediately after strand invasion, the Holliday junction moves along the linked DNA during the branch migration process. It is in this movement of the Holliday junction that base pairs between the two homologous DNA duplexes are exchanged. To catalyze branch migration, the RuvA protein first recognizes and binds to the Holliday junction and recruits the RuvB protein to form the RuvAB complex. Two sets of the RuvB protein, which each form a ring-shaped ATPase, are loaded onto opposite sides of the Holliday junction, where they act as twin pumps that provide the force for branch migration. Between those two rings of RuvB, two sets of the RuvA protein assemble in the center of the Holliday junction such that the DNA at the junction is sandwiched between each set of RuvA. The strands of both DNA duplexes—the "donor" and the "recipient" duplexes—are unwound on the surface of RuvA as they are guided by the protein from one duplex to the other.

Resolution

In the resolution phase of recombination, any Holliday junctions formed by the strand invasion process are cut, thereby restoring two separate DNA molecules. This cleavage is done by RuvAB complex interacting with RuvC, which together form the RuvABC complex. RuvC is an endonuclease that cuts the degenerate sequence 5'-(A/T)TT(G/C)-3'. The sequence is found frequently in DNA, about once every 64 nucleotides. Before cutting, RuvC likely gains access to the Holliday junction by displacing one of the two RuvA tetramers covering the DNA there. Recombination results in either "splice" or "patch" products, depending on how RuvC cleaves the Holliday junction. Splice products are crossover products, in which there is a rearrangement of genetic material around the site of recombination. Patch products, on the other hand, are non-crossover products in which there is no such rearrangement and there is only a "patch" of hybrid DNA in the recombination product.

Facilitating Genetic Transfer

Homologous recombination is an important method of integrating donor DNA into a recipient organism's genome in horizontal gene transfer, the process by which an organism incorporates foreign DNA from another organism without being the offspring of that organism. Homologous recombination requires incoming DNA to be highly similar to the recipient genome, and so horizontal gene transfer is usually limited to similar bacteria. Studies in several species of bacteria have established that there is a log-linear decrease in recombination frequency with increasing difference in sequence between host and recipient DNA.

In bacterial conjugation, where DNA is transferred between bacteria through direct cell-to-cell contact, homologous recombination helps integrate foreign DNA into the host genome via the RecBCD pathway. The RecBCD enzyme promotes recombination after DNA is converted from single-strand DNA–in which form it originally enters the bacterium–to double-strand DNA during replication. The RecBCD pathway is also essential for the final phase of transduction, a type of horizontal gene transfer in which DNA is transferred from one bacterium to another by a virus. Foreign, bacterial DNA is sometimes misincorporated in the capsid head of bacteriophage virus particles as DNA is packaged into new bacteriophages during viral replication. When these new bacteriophages infect other bacteria, DNA from the previous host bacterium is injected into the new bacterial host as double-strand DNA. The RecBCD enzyme then incorporates this double-strand DNA into the genome of the new bacterial host.

Bacterial Transformation

Natural bacterial transformation involves the transfer of DNA from a donor bacterium to a recipient bacterium, where both donor and recipient are ordinarily of the same species. Transformation, unlike bacterial conjugation and transduction, depends on numerous bacterial gene products that specifically interact to perform this process. Thus transformation is clearly a bacterial adaptation for DNA transfer. In order for a bacterium to bind, take up and integrate donor DNA into its resident chromosome by homologous recombination, it must first enter a special physiological state termed competence. The *RecA/Rad51/DMC1* gene family plays a central role in homologous recombination during bacterial transformation as it does during eukaryotic meiosis and mitosis. For instance, the RecA protein is essential for transformation in *Bacillus subtilis* and *Streptococcus pneumoniae*, and expression of the RecA gene is induced during the development of competence for transformation in these organisms.

As part of the transformation process, the RecA protein interacts with entering single-stranded DNA (ssDNA) to form RecA/ssDNA nucleofilaments that scan the resident chromosome for regions of homology and bring the entering ssDNA to the corresponding region, where strand exchange and homologous recombination occur. Thus the process of homologous recombination during bacterial transformation has fundamental similarities to homologous recombination during meiosis.

In Viruses

Homologous recombination occurs in several groups of viruses. In DNA viruses such as herpesvirus, recombination occurs through a break-and-rejoin mechanism like in bacteria and eukaryotes. There is also evidence for recombination in some RNA viruses, specifically positive-sense ssRNA viruses like retroviruses, picornaviruses, and coronaviruses. There is controversy over whether homologous recombination occurs in negative-sense ssRNA viruses like influenza.

In RNA viruses, homologous recombination can be either precise or imprecise. In the precise type of RNA-RNA recombination, there is no difference between the two parental RNA sequences and the resulting crossover RNA region. Because of this, it is often difficult to determine the location of crossover events between two recombining RNA sequences. In imprecise RNA homologous recombination, the crossover region has some difference with the parental RNA sequences – caused by either addition, deletion, or other modification of nucleotides. The level of precision in crossover is controlled by the sequence context of the two recombining strands of RNA: sequences rich in adenine and uracil decrease crossover precision.

Homologous recombination is important in facilitating viral evolution. For example, if the genomes of two viruses with different disadvantageous mutations undergo recombination, then they may be able to regenerate a fully functional genome. Alternatively, if two similar viruses have infected the same host cell, homologous recombination can allow those two viruses to swap genes and thereby evolve more potent variations of themselves.

Homologous recombination is the proposed mechanism whereby the DNA virus *human herpesvirus-6* integrates into human telomeres.

When two or more viruses, each containing lethal genomic damage, infect the same host cell, the virus genomes can often pair with each other and undergo homologous recombinational repair to produce viable progeny. This process, known as multiplicity reactivation, has been studied in several bacteriophages, including phage T4. Enzymes employed in recombinational repair in phage T4 are functionally homologous to enzymes employed in bacterial and eukaryotic recombinational repair. In particular, with regard to a gene necessary for the strand exchange reaction, a key step in homologous recombinational repair, there is functional homology from viruses to humans (i. e. *uvsX* in phage T4; *recA* in E. coli and other bacteria, and *rad51* and *dmc1* in yeast and other eukaryotes, including humans). Multiplicity reactivation has also been demonstrated in numerous pathogenic viruses.

Effects of Dysfunction

Without proper homologous recombination, chromosomes often incorrectly align for the first phase of cell division in meiosis. This causes chromosomes to fail to properly segregate in a process called nondisjunction. In turn, nondisjunction can cause sperm and ova to have too few or too many chromosomes. Down's syndrome, which is caused by an extra copy of chromosome 21, is one of many abnormalities that result from such a failure of homologous recombination in meiosis.

Deficiencies in homologous recombination have been strongly linked to cancer formation in humans. For example, each of the cancer-related diseases Bloom's syndrome, Werner's syndrome and Rothmund-Thomson syndrome are caused by malfunctioning copies of RecQ helicase genes involved in the regulation of homologous recombination: BLM, WRN and RECQ4, respectively. In the cells of Bloom's syndrome patients, who lack a working copy of the BLM protein, there is an elevated rate of homologous recombination. Experiments in mice deficient in BLM have suggested that the mutation gives rise to cancer through a loss of heterozygosity caused by increased homologous recombination. A loss in heterozygosity refers to the loss of one of two versions—or alleles—of a gene. If one of the lost alleles helps to suppress tumors, like the gene for the retinoblastoma protein for example, then the loss of heterozygosity can lead to cancer.

Decreased rates of homologous recombination cause inefficient DNA repair, which can also lead to cancer. This is the case with BRCA1 and BRCA2, two similar tumor suppressor genes whose malfunctioning has been linked with considerably increased risk for breast and ovarian cancer. Cells missing BRCA1 and BRCA2 have a decreased rate of homologous recombination and increased sensitivity to ionizing radiation, suggesting that decreased homologous recombination leads to increased susceptibility to cancer. Because the only known function of BRCA2 is to help initiate homologous recombination, researchers have speculated that more detailed knowledge of BRCA2's role in homologous recombination may be the key to understanding the causes of breast and ovarian cancer.

Evolutionary Conservation

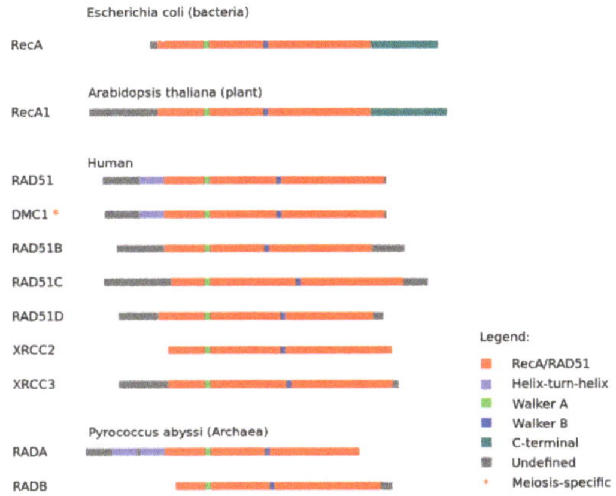

Figure. Protein domains in homologous recombination-related proteins are conserved across the three main groups of life: archaea, bacteria and eukaryotes.

While the pathways can mechanistically vary, the ability of organisms to perform homologous recombination is universally conserved across all domains of life. Based on the similarity of their amino acid sequences, homologs of a number of proteins can be found in multiple domains of life indicating that they evolved a long time ago, and have since diverged from common ancestral proteins.

RecA recombinase family members are found in almost all organisms with RecA in bacteria, Rad51 and DMC1 in eukaryotes, RadA in archaea, and UvsX in T4 phage.

Related single stranded binding proteins that are important for homologous recombination, and many other processes, are also found in all domains of life.

Rad54, Mre11, Rad50, and a number of other proteins are also found in both archaea and eukaryotes.

The RecA Recombinase Family

The proteins of the RecA recombinase family of proteins are thought to be descended from a common ancestral recombinase. The RecA recombinase family contains RecA protein from bacteria, the Rad51 and Dmc1 proteins from eukaryotes, and RadA from archaea, and the recombinase paralog proteins. Studies modeling the evolutionary relationships between the Rad51, Dmc1 and RadA proteins indicate that they are monophyletic, or that they share a common molecular ancestor. Within this protein family, Rad51 and Dmc1 are grouped together in a separate clade from RadA. One of the reasons for grouping these three proteins together is that they all possess a modified helix-turn-helix motif, which helps the proteins bind to DNA, toward their N-terminal ends. An ancient gene duplication event of a eukaryotic RecA gene and subsequent mutation has been proposed as a likely origin of the modern RAD51 and DMC1 genes.

The proteins generally share a long conserved region known as the RecA/Rad51 domain. Within this protein domain are two sequence motifs, Walker A motif and Walker B motif. The Walker A and B motifs allow members of the RecA/Rad51 protein family to engage in ATP binding and ATP hydrolysis.

Meiosis-specific Proteins

The discovery of Dmc1 in several species of *Giardia*, one of the earliest protists to diverge as a eukaryote, suggests that meiotic homologous recombination—and thus meiosis itself—emerged very early in eukaryotic evolution. In addition to research on Dmc1, studies on the Spo11 protein have provided information on the origins of meiotic recombination. Spo11, a type II topoisomerase, can initiate homologous recombination in meiosis by making targeted double-strand breaks in DNA. Phylogenetic trees based on the sequence of genes similar to SPO11 in animals, fungi, plants, protists and archaea have led scientists to believe that the version Spo11 currently in eukaryotes emerged in the last common ancestor of eukaryotes and archaea.

Technological Applications

Gene Targeting

Many methods for introducing DNA sequences into organisms to create recombinant DNA and genetically modified organisms use the process of homologous recombination. Also called gene targeting, the method is especially common in yeast and mouse genetics. The gene targeting method in knockout mice uses mouse embryonic stem cells to deliver artificial genetic material (mostly of therapeutic interest), which represses the target gene of the mouse by the principle of homologous recombination. The mouse thereby acts as a working model to understand the effects of a specific mammalian gene. In recognition of their discovery of how homologous recombination can be used to introduce genetic modifications in mice through embryonic stem cells, Mario Capecchi, Martin Evans and Oliver Smithies were awarded the 2007 Nobel Prize for Physiology or Medicine.

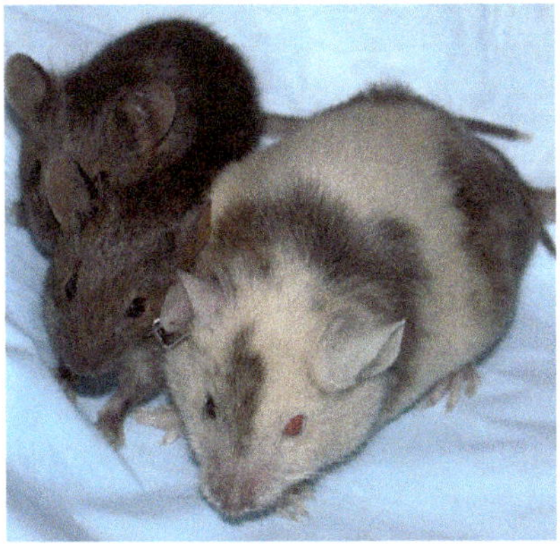

Figure. As a developing embryo, this chimeric mouse had the agouti coat color gene introduced into its DNA via gene targeting. Its offspring are homozygous for the agouti gene.

Advances in gene targeting technologies which hijack the homologous recombination mechanics of cells are now leading to the development of a new wave of more accurate, isogenic human disease models. These engineered human cell models are thought to more accurately reflect the genetics of human diseases than their mouse model predecessors. This is largely because mutations of interest are introduced into endogenous genes, just as they occur in the real patients, and because they are based on human genomes rather than rat genomes. Furthermore, certain technologies enable the knock-in of a particular mutation rather than just knock-outs associated with older gene targeting technologies.

Protein Engineering

Protein engineering with homologous recombination develops chimeric proteins by swapping fragments between two parental proteins. These techniques exploit the fact that recombination can introduce a high degree of sequence diversity while preserving a protein's ability to fold into its tertiary structure, or three-dimensional shape. This stands in contrast to other protein engineering techniques, like random point mutagenesis, in which the probability of maintaining protein function declines exponentially with increasing amino acid substitutions. The chimeras produced by recombination techniques are able to maintain their ability to fold because their swapped parental fragments are structurally and evolutionarily conserved. These recombinable "building blocks" preserve structurally important interactions like points of physical contact between different amino acids in the protein's structure. Computational methods like SCHEMA and statistical coupling analysis can be used to identify structural subunits suitable for recombination.

Techniques that rely on homologous recombination have been used to engineer new proteins. In a study published in 2007, researchers were able to create chimeras of two enzymes involved in the biosynthesis of isoprenoids, a diverse class of compounds including hormones, visual pigments and certain pheromones. The chimeric proteins acquired an ability to catalyze an essential reaction in isoprenoid biosynthesis—one of the most diverse pathways of biosynthesis found in nature—that was absent in the parent proteins. Protein engineering through recombination has also produced chimeric enzymes with new function in members of a group of proteins known as the cytochrome P450 family, which in humans is involved in detoxifying foreign compounds like drugs, food additives and preservatives.

Cancer Therapy

Cancer cells with BRCA mutations have deficiencies in homologous recombination, and drugs to exploit those deficiencies have been developed and used successfully in clinical trials. Olaparib, a PARP1 inhibitor, shrunk or stopped the growth of tumors from breast, ovarian and prostate cancers caused by mutations in the BRCA1 or BRCA2 genes, which are necessary for HR. When BRCA1 or BRCA2 is absent, other types of DNA repair mechanisms must compensate for the deficiency of HR, such as base-excision repair (BER) for stalled replication forks or non-homologous end joining (NHEJ) for double strand breaks. By inhibiting BER in an HR-deficient cell, olaparib applies the concept of synthetic lethality to specifically target cancer cells. While PARP1 inhibitors represent a novel approach to cancer therapy, researchers have cautioned that they may prove insufficient for treating late-stage metastatic cancers. Cancer cells can become resistant to a PARP1 inhibitor if they undergo deletions of mutations in BRCA2, undermining the drug's synthetic lethality by restoring cancer cells' ability to repair DNA by HR.

Synapsis

Synapsis (also called syndesis) is the pairing of two homologous chromosomes that occurs during meiosis. It allows matching-up of homologous pairs prior to their segregation, and possible chromosomal crossover between them. Synapsis takes place during prophase I of meiosis. When homologous chromosomes synapse, their ends are first attached to the nuclear envelope. These end-membrane complexes then migrate, assisted by the extranuclear cytoskeleton, until matching ends have been paired. Then the intervening regions of the chromosome are brought together, and may be connected by a protein-RNA complex called the synaptonemal complex. While autosomes undergo synapsis during meiosis, sex chromosomes often remain unpaired.

Mitosis also has prophase, but does not ordinarily do pairing of two homologous chromosomes.

When the non-sister chromatids intertwine, segments of chromatids with similar sequence may break apart and be exchanged in a process known as genetic recombination or "crossing-over". This exchange produces a chiasma, a region that is shaped like an X, where the two chromosomes are physically joined. At least one chiasma per chromosome often appears to be necessary to stabilise bivalents along the metaphase plate during separation. The crossover of genetic material also provides a possible defence against 'chromosome killer' mechanisms, by removing the distinction between 'self' and 'non-self' through which such a mechanism could operate. A further consequence of recombinant synapsis is to increase genetic variability within the offspring. Repeated recombination also has the general effect of allowing genes to move independently of each other through the generations, allowing for the independent concentration of beneficial genes and the purging of the detrimental.

Following synapsis, a type of recombination referred to as synthesis dependent strand annealing (SDSA) occurs frequently. SDSA recombination involves information exchange between paired non-sister homologous chromatids, but not physical exchange. SDSA recombination does not cause crossing-over. Both the non-crossover and crossover types of recombination function as processes for repairing DNA damage, particularly double-strand breaks.

The central function of synapsis is therefore the identification of homologues by pairing, an essential step for a successful meiosis. The processes of DNA repair and chiasma formation that take place following synapsis have consequences at many levels, from cellular survival through to impacts upon evolution itself.

Spindle Apparatus

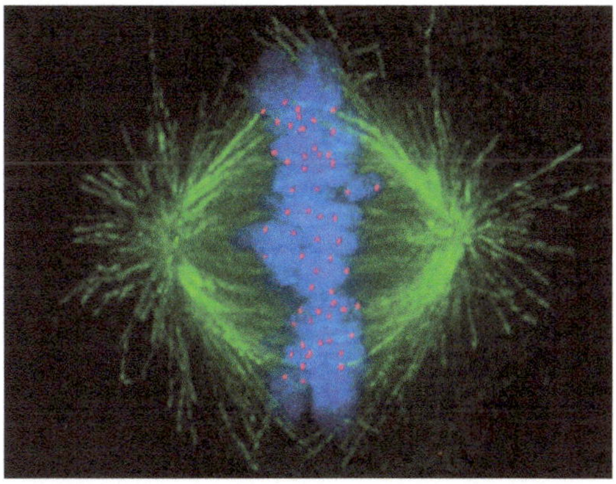

Micrograph showing condensed chromosomes in blue, kinetochores in pink, and microtubules in green during metaphase of mitosis

In cell biology, the spindle apparatus refers to the cytoskeletal structure of eukaryotic cells that forms during cell division to separate sister chromatids between daughter cells. It is referred to as the mitotic spindle during mitosis, a process that produces genetically identical daughter cells, or the meiotic spindle during meiosis, a process that produces gametes with half the number of chromosomes of the parent cell.

Besides chromosomes, the spindle apparatus is composed of hundreds of proteins. Microtubules comprise the most abundant components of the machinery.

Spindle Structure

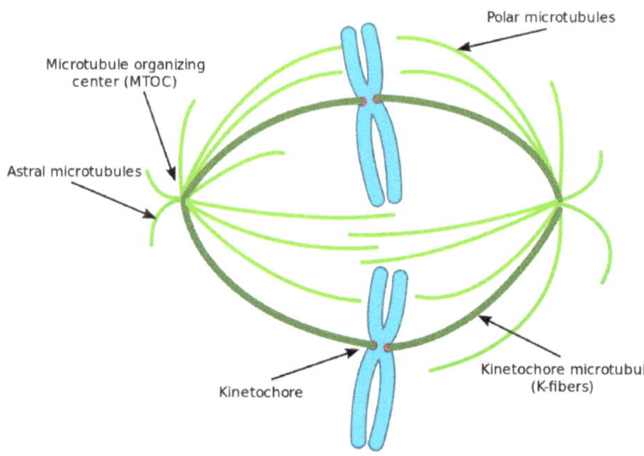

This diagram depicts the organization of a typical mitotic spindle found in animal cells. Chromosomes are attached to kinetochore microtubules via a multiprotein complex called the kinetochore. Polar microtubules interdigitate at the spindle midzone and push the spindle poles apart via motor proteins. Astral microtubules anchor the spindle poles to the cell membrane. Microtubule polymerization is nucleated at the microtubule organizing center.

Attachment of microtubules to chromosomes is mediated by kinetochores, which actively monitor spindle formation and prevent premature anaphase onset. Microtubule polymerization and depolymerization dynamics drive chromosome congression. Depolymerization of microtubules generates tension at kinetochores; bipolar attachment of sister kinetochores to microtubules emanating from opposite cell poles couples opposing tension forces, aligning chromosomes at the cell equator and poising them for segregation to daughter cells. Once every chromosome is bi-oriented, anaphase commences and cohesin, which couples sister chromatids, is severed, permitting the transit of the sister chromatids to opposite poles.

The cellular spindle apparatus includes the spindle microtubules, associated proteins, which include kinesin and dynein molecular motors, condensed chromosomes, and any centrosomes or asters that may be present at the spindle poles depending on the cell type. The spindle apparatus is vaguely ellipsoid in cross section and tapers at the ends. In the wide middle portion, known as the spindle midzone, antiparallel microtubules are bundled by kinesins. At the pointed ends, known as spindle poles, microtubules are nucleated by the centrosomes in most animal cells. *Acentrosomal* or *anastral* spindles lack centrosomes or asters at the spindle poles, respectively, and occur for example during female meiosis in most animals. In this instance, a Ran GTP gradient is the main regulator of spindle microtubule organization and assembly. In fungi, spindles form between spindle pole bodies embedded in the nuclear envelope, which does not break down during mitosis.

Microtubule-associated Proteins and Spindle Dynamics

The dynamic lengthening and shortening of spindle microtubules, through a process known as dynamic instability determines to a large extent the shape of the mitotic spindle and promotes the proper alignment of chromosomes at the spindle midzone. Microtubule-associated proteins (MAPs) associate with microtubules at the midzone and the spindle poles to regulate their dynamics. γ-tubulin is a specialized tubulin variant that assembles into a ring complex called γ-TuRC which nucleates polymerization of α/β tubulin heterodimers into microtubules. Recruitment of γ-TuRC to the pericentrosomal region stabilizes microtubule minus-ends and anchors them near the microtubule-organizing center.The microtubule-associated protein Augmin acts in conjunction with γ-TURC to nucleate new microtubules off of existing microtubules.

The growing ends of microtubules are protected against catastrophe by the action of plus-end microtubule tracking proteins (+TIPs) to promote their association with kinetochores at the midzone. CLIP170 was shown to localize near microtubule plus-ends in HeLa cells and to accumulate in kinetochores during prometaphase. Although how CLIP170 recognizes plus-ends remains unclear, it has been shown that its homologues protect against catastrophe and promote rescue, suggesting a role for CLIP170 in stabilizing plus-ends and possibly mediating their direct attachment to kinetochores. CLIP-associated proteins like CLASP1 in humans have also been shown to localize to plus-ends and the outer kinetochore as well as to modulate the dynamics of kinetochore microtubules (Maiato 2003). CLASP homologues in *Drosophila*, *Xenopus*, and yeast are required for proper spindle assembly; in mammals, CLASP1 and CLASP2 both contribute to proper spindle assembly and microtubule dynamics in anaphase. Plus-end polymerization may be further moderated by the EB1 protein, which directly binds the growing ends of microtubules and coordinates the binding of other +TIPs.

Opposing the action of these microtubule-stabilizing proteins are a number of microtubule-depolymerizing factors which permit the dynamic remodeling of the mitotic spindle to promote chromosome congression and attainment of bipolarity. The kinesin-13 superfamily of MAPs contains a class of plus-end-directed motor proteins with associated microtubule depolymerization activity including the well-studied mammalian MCAK and *Xenopus* XKCM1. MCAK localizes to the growing tips of microtubules at kinetochores where it can trigger catastrophe in direct competition with stabilizing +TIP activity. These proteins harness the energy of ATP hydrolysis to induce destabilizing conformational changes in protofilament structure that cause kinesin release and microtubule depolymerization. Loss of their activity results in numerous mitotic defects. Additional microtubule destabilizing proteins include Op18/stathmin and katanin which have roles in remodeling the mitotic spindle as well as promoting chromosome segregation during anaphase.

The activities of these MAPs are carefully regulated to maintain proper microtubule dynamics during spindle assembly, with many of these proteins serving as Aurora and Polo-like kinase substrates.

Organizing the Spindle Apparatus

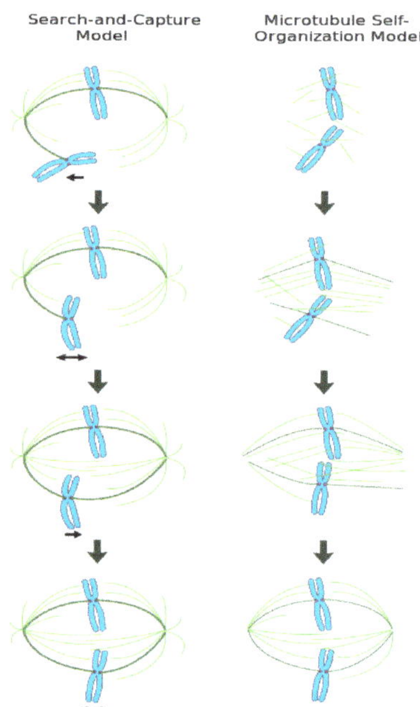

In the centrosome-mediated "search and capture" model (left), microtubules nucleated from centrosomes contact chromosomes by chance and become stabilized at kinetochores to form the spindle. In the chromatin-mediated "self-organization" model (right), microtubules are nucleated around the vicinity of mitotic chromatin and organized into a bipolar array by motor proteins.

In a properly formed mitotic spindle, bi-oriented chromosomes are aligned along the equator of the cell with spindle microtubules oriented roughly perpendicular to the chromosomes, their plus-ends embedded in kinetochores and their minus-ends anchored at the cell poles. The precise orientation of this complex is required to ensure accurate chromosome segregation and to specify the cell division plane. However, it remains unclear how the spindle becomes organized. Two models predominate the field, which are synergistic and not mutually exclusive. In the *search-and-capture model*, the spindle is predominantly organized by the poleward separation of centrosomal microtubule organizing centers (MTOCs). Spindle microtubules emanate from centrosomes and 'seek' out kinetochores; when they bind a kinetochore they become stabilized and exert tension on the chromosomes. In an alternative *self assembly* model, microtubules undergo acentrosomal nucleation among the condensed chromosomes.

Constrained by cellular dimensions, lateral associations with antiparallel microtubules via motor proteins, and end-on attachments to kinetochores, microtubules naturally adopt a spindle-like structure with chromosomes aligned along the cell equator.

Centrosome-mediated "Search-and-capture" Model

In this model, microtubules are nucleated at microtubule organizing centers and undergo rapid growth and catastrophe to 'search' the cytoplasm for kinetochores. Once they bind a kinetochore, they are stabilized and their dynamics are reduced. The newly mono-oriented chromosome oscillates in space near the pole to which it is attached until a microtubule from the opposite pole binds the sister kinetochore. This second attachment further stabilizes kinetochore attachment to the mitotic spindle. Gradually, the bi-oriented chromosome is pulled towards the center of the cell until microtubule tension is balanced on both sides of the centromere; the congressed chromosome then oscillates at the metaphase plate until anaphase onset releases cohesion of the sister chromatids.

In this model, microtubule organizing centers are localized to the cell poles, their separation driven by microtubule polymerization and 'sliding' of antiparallel spindle microtubules with respect to one another at the spindle midzone mediated by bipolar, plus-end-directed kinesins. Such sliding forces may not only account for spindle pole separation early in mitosis, but also spindle elongation during late anaphase.

Chromatin-mediated Self-organization of the Mitotic Spindle

In contrast to the search-and-capture mechanism in which centrosomes largely dictate the organization of the mitotic spindle, this model proposes that microtubules are nucleated acentrosomally near chromosomes and spontaneously assemble into anti-parallel bundles and adopt a spindle-like structure. Classic experiments by Heald and Karsenti show that functional mitotic spindles and nuclei form around DNA-coated beads incubated in *Xenopus* egg extracts and that bipolar arrays of microtubules are formed in the absence of centrosomes and kinetochores. Indeed, it has also been shown that laser ablation of centrosomes in vertebrate cells inhibits neither spindle assembly nor chromosome segregation. Under this scheme, the shape and size of the mitotic spindle are a function of the biophysical properties of the cross-linking motor proteins.

Chromatin-mediated microtubule nucleation by the Ran GTP gradient

The guanine nucleotide exchange factor for the small GTPase Ran (Regulator of chromosome condensation 1 or RCC1) is attached to nucleosomes via core histones H2A and H2B. Thus, a gradient of GTP-bound Ran is generated around the vicinity of mitotic chromatin. Glass beads coated with RCC1 induce microtubule nucleation and bipolar spindle formation in *Xenopus* egg extracts, revealing that the Ran GTP gradient alone is sufficient for spindle assembly. The gradient triggers release of spindle assembly factors (SAFs) from inhibitory interactions via the transport proteins importin β/α. The unbound SAFs then promote microtubule nucleation and stabilization around mitotic chromatin, and spindle bipolarity is organized by microtubule motor proteins.

Regulation of Spindle Assembly

Spindle assembly is largely regulated by phosphorylation events catalyzed by mitotic kinases. Cyclin dependent kinase complexes (CDKs) are activated by mitotic cyclins, whose translation increases during mitosis. CDK1 (also called CDC2) is considered the main mitotic kinase in mammalian cells and is activated by Cyclin B1. Aurora kinases are required for proper spindle assembly and separation. Aurora A associates with centrosomes and is believed to regulate mitotic entry. Aurora B is a member of the chromosomal passenger complex and mediates chromosome-microtubule attachment and sister chromatid cohesion. Polo-like kinase, also known as PLK, especially PLK1 has important roles in the spindle maintenance by regulating microtubule dynamics.

Mitotic Chromosome Structure

By the end of DNA replication, sister chromatids are bound together in an amorphous mass of tangled DNA and protein that would be virtually impossible to partition into each daughter cell. To avoid this problem, mitotic entry triggers a dramatic reorganization of the duplicated genome. Sister chromatids are disentangled and resolved from one another. Chromosomes also shorten in length, up to 10,000 fold in animal cells, in a process called condensation. Condensation begins in prophase and chromosomes are maximally compacted into rod-shaped structures by the time they are aligned in the middle of the spindle at metaphase. This gives mitotic chromosomes the classic "X" shape seen in karyotypes, with each condensed sister chromatid linked along their lengths by cohesin proteins and joined, often near the center, at the centromere.

While these dynamic rearrangements are vitally important to ensure accurate and high-fidelity segregation of the genome, our understanding of mitotic chromosome structure remains largely incomplete. A few specific molecular players have been identified, however: Topoisomerase II uses ATP hydrolysis to catalyze decatenation of DNA entanglements, promoting sister chromatid resolution. Condensins are 5-subunit complexes that also use ATP-hydrolysis to promote chromosome condensation. Experiments in *Xenopus* egg extracts have also implicated linker Histone H1 as an important regulator of mitotic chromosome compaction.

Mitotic Spindle Assembly Checkpoint

The completion of spindle formation is a crucial transition point in the cell cycle called the spindle assembly checkpoint. If chromosomes are not properly attached to the mitotic spindle by the time of this checkpoint, the onset of anaphase will be delayed. Failure of this spindle assembly checkpoint can result in aneuploidy and may be involved in aging and the formation of cancer.

Spindle Apparatus Orientation

Cell division orientation is of the major importance for tissue architecture, cell fates and morphogenesis. Cells tend to divide along their long axis according of so called Hertwig rule. The axis of cell division is determined by the orientation of the spindle apparatus. Cells divide along the line connecting two centrosomes of spindle apparatus. After formation, the spindle apparatus undergoes rotation inside the cell. The astral microtubules, originating from centrosomes reach the cell membrane where they are pulled towards specific cortical clues. *In vitro*, the distribution of cortical clues set up by the adhesive pattern. *In vivo* polarity cues are determined by localization of Tricellular junctions localized at cell vertices. The spatial distribution of cortical clues leads to the force field that determine final spingle apparatus orientation and following orientation of cell division.

Synaptonemal Complex

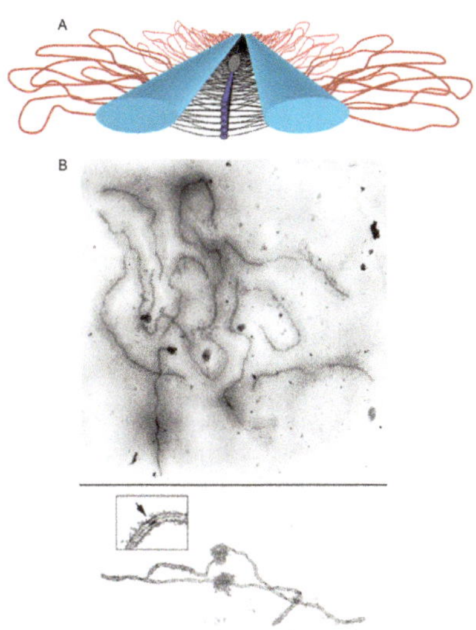

A Homologous chromosomes (light blue) align and synapse together via transverse filaments (black lines) and longitudinal filaments (dark blue). Recombination nodules (gray ellipsoids) on the central region may help in completing recombination. Chromatin (red loops) is attached to its corresponding chromosome, extending from both sister chromatids. B Top: Set of tomato SCs. Chromatin "sheaths" visible around each SC. Bottom: Two tomato SCs with the chromatin removed, allowing kinetochores ("ball-like" structures) at centromeres to be revealed.

The synaptonemal complex is a protein structure that forms between homologous chromosomes (two pairs of sister chromatids) during meiosis and is thought to mediate chromosome pairing, synapsis, and recombination. It is now evident that the synaptonemal complex is not required for genetic recombination. Research has shown that not only does it form after genetic recombination but mutant yeast cells unable to assemble a synaptonemal complex can still engage in the exchange of genetic information. It is currently thought that the SC functions primarily as a scaffold to allow interacting chromatids to complete their crossover activities. The synaptonemal complex is a tripartite structure consisting of two parallel lateral regions and a central element.

Three specific components of the synaptonemal complex have been characterized: SC protein-1 (SYCP1), SC protein-2 (SYCP2), and SC protein-3 (SYCP3). In humans, the SYCP1 gene is on chromosome 1p13; the SYCP2 gene is on chromosome 20q13.33; and the gene for SYCP3 is on chromosome 12q.

The synaptonemal complex (SC) was described by Montrose J. Moses in 1956 in primary spermatocytes of crayfish and by D. Fawcett in spermatocytes of pigeon, cat and man. As seen with the electron microscope, the synaptonemal complex is formed by two "lateral elements", mainly formed by SYCP3 and secondarily by SYCP2, a "central element" that contains at least two additional proteins and the amino terminal region of SYCP1, and a "central region" spanned between the two lateral elements, that contains the "transverse filaments" composed mainly by the protein SYCP1.

The SCs can be seen with the light microscope using silver staining or with immunofluorescence techniques that label the proteins SYCP3 or SYCP2.

This "tripartite structure" is seen during the pachytene stage of the first meiotic prophase, both in males and in females during gametogenesis. Previous to the pachytene stage, during leptonema, the lateral elements begin to form and they initiate and complete their pairing during the zygotene stage. After pachynema ends, the SC usually becomes disassembled and can no longer be identified.

Formation of the SC usually reflects the pairing or "synapsis" of homologous chromosomes and may be used to probe the presence of pairing abnormalities in individuals carrying chromosomal abnormalities, either in number or in the chromosomal structure. The sex chromosomes in male mammals show only "partial synapsis" as they usually form only a short SC in the XY pair. The SC shows very little structural variability among eukaryotic organisms despite some significant protein differences. In many organisms the SC carries one or several "recombination nodules" associated to its central space. These nodules are thought to correspond to mature genetic recombination events or "crossovers".

In cell development the synaptonemal complex disappears during the late prophase of meiosis I.

One aspect of meiosis research (in genetics, molecular biology, and reproductive medicine) involving synaptonemal complexes is how they (or what they do) can be affected and damaged by chemical exposure, specifically by toxins like bisphenol A (an experiment at Harvard Medical School by Dr. Monica Colaiácovo, Ph.D., an associate professor of genetics, involved dosing C. elegans worms with that toxin and then verifying the results in mouse studies).

Mitotic Index

Mitotic index is defined as the ratio between the number of cells in a population undergoing mitosis to the number of cells.

Purpose

The purpose of the mitotic index is to measure cellular proliferation. The mitotic index is an important prognostic factor predicting both overall survival and response to chemotherapy in most types of cancer. It may lose much of its predictive value for elderly populations, for example a low mitotic index loses any prognostic value for women over 70 years old with breast cancer.

Additionally, the mitotic index is simply a measurement to determine the percentage of cells undergoing mitosis. Mitosis is the division of somatic cells when genetic information from one single cell is equally dispersed into two daughter cells. Durations of the cell cycle and mitosis vary in different cell types. An elevated mitotic index indicates more cells are dividing, and thus obvious in cancer cells, The mitotic index may be elevated during necessary processes to life, such as the normal growth of plants or animals, as well as cellular repair the site of an injury.

Calculation

Mitotic index can be calculated using the following operation : cells observed with visible chromosomes ÷ total number of cells visible.

A typical figure of mitotic index includes statements like "10 mitotic figures are noted per 10 high power fields" followed by "4 mitotic figures noted per 50 high power fields". Mitotic index: 14 MF/10 400x fields

Formula

(P+M+A+T) — the sum of all cells in phase as prophase, metaphase, anaphase and telophase, respectively; N — total number of cells.

References

- Eichorn, Susan, et al. Esau's Plant Anatomy: Meristems, Cells, and Tissues of the Plant Body: Their Structure, Function, and Development, 3rd Edition. 2006. ISBN 978-0-471-73843-5
- Imperfect DNA replication results in mutations. Berg JM, Tymoczko JL, Stryer L, Clarke ND (2002). "Chapter 27: DNA Replication, Recombination, and Repair". Biochemistry. W.H.

Freeman and Company. ISBN 0-7167-3051-0.

- Alberts B, Johnson A, Lewis J, Raff M, Roberts K, Walter P (2002). "5DNA Replication, Repair, and Recombination". Molecular Biology of the Cell. Garland Science. ISBN 0-8153-3218-1.
- Berg JM, Tymoczko JL, Stryer L, Clarke ND (2002). "Chapter 27, Section 4: DNA Replication of Both Strands Proceeds Rapidly from Specific Start Sites". Biochemistry. W.H. Freeman and Company. ISBN 0-7167-3051-0.
- Lodish H, Berk A, Zipursky LS, Matsudaira P, Baltimore D, Darnell J (2000). Molecular Cell Biology. W. H. Freeman and Company. ISBN 0-7167-3136-3.12.1.
- Griffiths A, et al. (1999). "8: Chromosome Mutations: Chromosomal Rearrangements". Modern Genetic Analysis. W. H. Freeman and Company. ISBN 0-7167-3118-5.
- Watson JD, Baker TA, Bell SP, Gann A, Levine M, Losick R (2003). Molecular Biology of the Gene (5th ed.). Pearson/Benjamin Cummings. pp. 259–291. ISBN 978-0-8053-4635-0.
- Fleischmann Jr WR (1996). "43". Medical Microbiology (4th ed.). University of Texas Medical Branch at Galveston. ISBN 0-9631172-1-1.
- Bernstein C, Bernstein H (2001). DNA repair in bacteriophage. In: Nickoloff JA, Hoekstra MF (Eds.) DNA Damage and Repair, Vol.3. Advances from Phage to Humans. Humana Press, Totowa, NJ, pp. 1–19. ISBN 978-0896038035
- Alberts B, Johnson A, Lewis J, Raff M, Roberts K, Walter P (2007). Molecular Biology of the Cell (5th ed.). Garland Science. ISBN 978-0-8153-4110-9.
- Morgan DO: The Cell Cycle: Principles of Control (Primers inBiology) London: New Science Press Ltd; 2007:297. ISBN 978-0-9539181-2-6
- Raven, Peter H.; Ray F. Evert; Susan E. Eichhorn (2005). Biology of Plants, 7th Edition. New York: W.H. Freeman and Company Publishers. p. 59. ISBN 0-7167-1007-2.

3

Occurrence of Meiosis in Eukaryotic Life Cycle

An organism whose cells contain a nucleus within its membrane is a eukaryote. Reproduction in eukaryotes can occur both sexually and asexually. While it happens asexually in mitosis, it occurs sexually in meiosis. This chapter on meiosis seeks to explain the development of sexual reproduction and the specific evolutionary characteristics that were propagated because of the same.

Sexual Reproduction

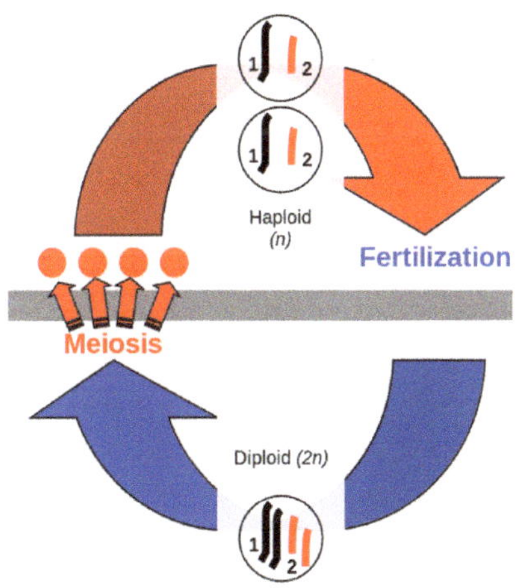

In the first stage of sexual reproduction, "meiosis", the number of chromosomes is reduced from a diploid number (2n) to a haploid number (n). During "fertilization", haploid gametes come together to form a diploid zygote and the original number of chromosomes is restored.

Sexual reproduction is a form of reproduction where two morphologically distinct types of specialized reproductive cells called gametes fuse together, involving a female's large ovum (or egg) and a male's smaller sperm. Each gamete contains half the number of chromosomes of normal cells. They are created by a specialized type of cell division, which only occurs in eukaryotic cells, known as meiosis. The two gametes fuse during fertilization to produce DNA replication and the creation of a single-celled zygote which includes genetic material from both gametes. In a process called genetic recombination, genetic material (DNA) joins up so that homologous chromosome sequences are aligned with each other, and this is followed by exchange of genetic information.

Two rounds of cell division then produce four daughter cells with half the number of chromosomes from each original parent cell, and the same number of chromosomes as both parents, though self-fertilization can occur. For instance, in human reproduction each human cell contains 46 chromosomes, 23 pairs, except gamete cells, which only contain 23 chromosomes, so the child will have 23 chromosomes from each parent genetically recombined into 23 pairs. Cell division initiates the development of a new individual organism in multicellular organisms, including animals and plants, for the vast majority of whom this is the primary method of reproduction. A species is defined as a taxonomic rank. A species is often defined as the largest group of organisms where two hybrids are capable of reproducing fertile offspring, typically using sexual reproduction, although the species problem encompasses a series of difficult related questions that often come up when biologists define the word species.

The evolution of sexual reproduction is a major puzzle because asexual reproduction should be able to outcompete it as every young organism created can bear its own young. This implies that an asexual population has an intrinsic capacity to grow more rapidly with each generation. This 50% cost is a fitness disadvantage of sexual reproduction. The two-fold cost of sex includes this cost and the fact that any organism can only pass on 50% of its own genes to its offspring. One definite advantage of sexual reproduction is that it prevents the accumulation of genetic mutations.

Sexual selection is a mode of natural selection in which some individuals out-reproduce others of a population because they are better at securing mates for sexual reproduction. It has been described as "a powerful evolutionary force that does not exist in asexual populations."

Prokaryotes, whose initial cell has additional or transformed genetic material, reproduce through asexual reproduction but may, in lateral gene transfer, display processes such as bacterial conjugation, transformation and transduction, which are similar to sexual reproduction although they do not lead to reproduction.

Evolution

The first fossilized evidence of sexual reproduction in eukaryotes is from the Stenian period, about 1 to 1.2 billion years ago.

Biologists studying evolution propose several explanations for why sexual reproduction developed and why it is maintained. These reasons include fighting the accumulation of deleterious mutations, increasing rate of adaptation to changing environments, dealing with competition, or masking deleterious mutations. All of these ideas about why sexual reproduction has been maintained are generally supported, but ultimately the size of the population determines if sexual reproduction is entirely beneficial. Larger populations appear to respond more quickly to benefits obtained through sexual reproduction than do smaller population sizes.

Maintenance of sexual reproduction has been explained by theories that work at several levels of selection, though some of these models remain controversial.

New models presented in recent years suggest a basic advantage for sexual reproduction in slowly reproducing complex organisms. Sexual reproduction allows these species to exhibit characteristics that depend on the specific environment that they inhabit, and the particular survival strategies that they employ.

Sexual Selection

In order to sexually reproduce, both males and females need to find a mate. Generally in animals mate choice is made by females while males compete to be chosen. This can lead organisms to extreme efforts in order to reproduce, such as combat and display, or produce extreme features caused by a positive feedback known as a Fisherian runaway. Thus sexual reproduction, as a form of natural selection, has an effect on evolution. Sexual dimorphism is where the basic phenotypic traits vary between males and females of the same species. Dimorphism is found in both sex organs and in secondary sex characteristics, body size, physical strength and morphology, biological ornamentation, behavior and other bodily traits. However, sexual selection is only implied over an extended period of time leading to sexual dimorphism.

Sex Ratio

Apart from some eusocial wasps, organisms which reproduce sexually have a 1:1 sex ratio of male and female births. The English statistician and biologist Ronald Fisher outlined why this is so in what has come to be known as Fisher's principle. This essentially says the following:

1. Suppose male births are less common than female.
2. A newborn male then has better mating prospects than a newborn female, and therefore can expect to have more offspring.
3. Therefore parents genetically disposed to produce males tend to have more than average numbers of grandchildren born to them.
4. Therefore the genes for male-producing tendencies spread, and male births become more common.
5. As the 1:1 sex ratio is approached, the advantage associated with producing males dies away.
6. The same reasoning holds if females are substituted for males throughout. Therefore 1:1 is the equilibrium ratio.

Animals

Insects

Australian emperor laying egg, guarded by the male

Insect species make up more than two-thirds of all extant animal species. Most insect species reproduce sexually, though some species are facultatively parthenogenetic. Many insects species have sexual dimorphism, while in others the sexes look nearly identical. Typically they have two sexes with males producing spermatozoa and females ova. The ova develop into eggs that have a covering called the chorion, which forms before internal fertilization. Insects have very diverse mating and reproductive strategies most often resulting in the male depositing spermatophore within the female, which she stores until she is ready for egg fertilization. After fertilization, and the formation of a zygote, and varying degrees of development, in many species the eggs are deposited outside the female; while in others, they develop further within the female and are born live.

Birds

Mammals

There are three extant kinds of mammals: monotremes, placentals and marsupials, all with internal fertilization. In placental mammals, offspring are born as juveniles: complete animals with the sex organs present although not reproductively functional. After several months or years, depending on the species, the sex organs develop further to maturity and the animal becomes sexually mature. Most female mammals are only fertile during certain periods during their estrous cycle, at which point they are ready to mate. Individual male and female mammals meet and carry out copulation.For most mammals, males and females exchange sexual partners throughout their adult lives.

Fish

The vast majority of fish species lay eggs that are then fertilized by the male, some species lay their eggs on a substrate like a rock or on plants, while others scatter their eggs and the eggs are fertilized as they drift or sink in the water column.

Some fish species use internal fertilization and then disperse the developing eggs or give birth to live offspring. Fish that have live-bearing offspring include the guppy and mollies or *Poecilia*. Fishes that give birth to live young can be ovoviviparous, where the eggs are fertilized within the female and the eggs simply hatch within the female body, or in seahorses, the male carries the developing young within a pouch, and gives birth to live young. Fishes can also be viviparous, where the female supplies nourishment to the internally growing offspring. Some fish are hermaphrodites, where a single fish is both male and female and can produce eggs and sperm. In hermaphroditic fish, some are male and female at the same time while in other fish they are serially hermaphroditic; starting as one sex and changing to the other. In at least one hermaphroditic species, self-fertilization occurs when the eggs and sperm are released together. Internal self-fertilization may occur in some other species. One fish species does not reproduce by sexual reproduction but uses sex to produce offspring; *Poecilia formosa* is a unisex species that uses a form of parthenogenesis called gynogenesis, where unfertilized eggs develop into embryos that produce female offspring. *Poecilia formosa* mate with males of other fish species that use internal fertilization, the sperm does not fertilize the eggs but stimulates the growth of the eggs which develops into embryos.

Plants

Animals typically produce gametes directly by meiosis. Male gametes are called sperm, and female gametes are called eggs or ova. In animals, fertilization follows immediately after meiosis. Plants on the other hand have mitosis occurring in spores, which are produced by meiosis. The spores germinate into the gametophyte phase. The gametophytes of different groups of plants vary in size; angiosperms have as few as three cells in pollen, and mosses and other so called primitive plants may have several million cells. Plants have an alternation of generations where the sporophyte phase is succeeded by the gametophyte phase. The sporophyte phase produces spores within the sporangium by meiosis.

Flowering Plants

Flowering plants are the dominant plant form on land and they reproduce either sexually or asexually. Often their most distinguishing feature is their reproductive organs, commonly called flowers. The anther produces pollen grains which contain the male gametophytes (sperm). For pollination to occur, pollen grains must attach to the stigma of the female reproductive structure (carpel), where the female gametophytes (ovules) are located inside the ovary. After the pollen tube grows through the carpel's style, the sex cell nuclei from the pollen grain migrate into the ovule to fertilize the egg cell and endosperm nuclei within the female gametophyte in a process termed double fertilization. The resulting zygote develops into an embryo, while the triploid endosperm (one sperm cell plus two female cells) and female tissues of the ovule give rise to the surrounding tissues in the developing seed. The ovary, which produced the female gametophyte(s), then grows into a fruit, which surrounds the seed(s). Plants may either self-pollinate or cross-pollinate.

Flowers are the sexual organs of flowering plants.

Nonflowering plants like ferns, moss and liverworts use other means of sexual reproduction.

In 2013, flowers dating from the Cretaceous (100 million years before present) were found encased in amber, the oldest evidence of sexual reproduction in a flowering plant. Microscopic images showed tubes growing out of pollen and penetrating the flower's stigma. The pollen was sticky, suggesting it was carried by insects.

Ferns

Ferns mostly produce large diploid sporophytes with rhizomes, roots and leaves; and on fertile leaves called sporangium, spores are produced. The spores are released and germinate to produce short, thin gametophytes that are typically heart shaped, small and green in color. The gametophytes or thallus, produce both motile sperm in the antheridia and egg cells in separate archegonia. After rains or when dew deposits a film of water, the motile sperm are splashed away from the antheridia, which are normally produced on the top side of the thallus, and swim in the film of water to the archegonia where they fertilize the egg. To promote out crossing or cross fertilization the sperm are released before the eggs are receptive of the sperm, making it more likely that the sperm will fertilize the eggs of different thallus. A zygote is formed after fertilization, which grows into a new sporophytic plant. The condition of having separate sporephyte and gametophyte plants is called alternation of generations. Other plants with similar reproductive means include the *Psilotum*, *Lycopodium*, *Selaginella* and *Equisetum*.

Bryophytes

The bryophytes, which include liverworts, hornworts and mosses, reproduce both sexually and vegetatively. They are small plants found growing in moist locations and like ferns, have motile sperm with flagella and need water to facilitate sexual reproduction. These plants start as a haploid spore that grows into the dominate form, which is a multicellular haploid body with leaf-like structures that photosynthesize. Haploid gametes are produced in antherida and archegonia by mitosis. The sperm released from the antherida respond to chemicals released by ripe archegonia and swim to them in a film of water and fertilize the egg cells thus producing a zygote. The zygote divides by mitotic division and grows into a sporophyte that is diploid. The multicellular diploid sporophyte produces structures called spore capsules, which are connected by seta to the archegonia. The spore capsules produce spores by meiosis, when ripe the capsules burst open and the spores are released. Bryophytes show considerable variation in their breeding structures and the above is a basic outline. Also in some species each plant is one sex while other species produce both sexes on the same plant.

Fungi

Fungi are classified by the methods of sexual reproduction they employ. The outcome of sexual reproduction most often is the production of resting spores that are used to survive inclement times and to spread. There are typically three phases in the sexual reproduction of fungi: plasmogamy, karyogamy and meiosis.

Bacteria and Archaea

Three distinct processes in prokaryotes are regarded as similar to eukaryotic sex: bacterial transformation, which involves the incorporation of foreign DNA into the bacterial chromosome; bacterial conjugation, which is a transfer of plasmid DNA between bacteria, but the plasmids are rarely incorporated into the bacterial chromosome; and gene transfer and genetic exchange in archaea.

Bacterial transformation involves the recombination of genetic material and its function is mainly associated with DNA repair. Bacterial transformation is a complex process encoded by numerous bacterial genes, and is a bacterial adaptation for DNA transfer. This process occurs naturally in at least 40 bacterial species. For a bacterium to bind, take up, and recombine exogenous DNA into its chromosome, it must enter a special physiological state referred to as competence. Sexual reproduction in early single-celled eukaryotes may have evolved from bacterial transformation, or from a similar process in archaea.

On the other hand, bacterial conjugation is a type of direct transfer of DNA between two bacteria through an external appendage called the conjugation pilus. Bacterial conjugation is controlled by plasmid genes that are adapted for spreading copies of the plasmid between bacteria. The infrequent integration of a plasmid into a host bacterial chromosome, and the subsequent transfer of a part of the host chromosome to another cell do not appear to be bacterial adaptations.

Exposure of hyperthermophilic archaeal Sulfolobus species to DNA damaging conditions induces cellular aggregation accompanied by high frequency genetic marker exchange. Ajon et al. hypothesized that this cellular aggregation enhances species-specific DNA repair by homologous recombination. DNA transfer in Sulfolobus may be an early form of sexual interaction similar to the more well-studied bacterial transformation systems that also involve species-specific DNA transfer leading to homologous recombinational repair of DNA damage.

Origin and Function of Meiosis

The origin and function of meiosis are fundamental to understanding the evolution of sexual reproduction in Eukaryotes. There is no current consensus among biologists on the questions of how sex in Eukaryotes arose in evolution, what basic function sexual reproduction serves, and why it is maintained, given the basic two-fold cost of sex. It is clear that it evolved over 1.2 billion years ago, and that almost all species which are descendents of the original sexually reproducing species are still sexual reproducers, including plants, fungi, and animals.

Meiosis is a key event of the sexual cycle in Eukaryotes. It is the stage of the life cycle when a cell gives rise to two haploid cells (gametes) each having half as many chromosomes. Two such haploid gametes, arising from different individual organisms, fuse by the process of fertilization, thus completing the sexual cycle.

Meiosis is ubiquitous among eukaryotes. It occurs in single-celled organisms such as yeast, as well as in multicellular organisms, such as humans. Eukaryotes arose from prokaryotes more than 1.5 billion years ago, and the earliest eukaryotes were likely single-celled organisms. To understand sex in eukaryotes, it is necessary to understand (1) how meiosis arose in single celled eukaryotes, and (2) the function of meiosis.

Origin of Meiosis

There are two conflicting theories on how meiosis arose. One is that meiosis evolved from prokaryotic sex as eukaryotes evolved from prokaryotes. The other is that meiosis arose from mitosis.

Theory that Meiosis Evolved from Prokaryotic Sex

In prokaryotic sex, DNA from one prokaryote is released into the surrounding medium, is then taken up by another prokaryote and its information integrated into the DNA of the recipient prokaryote. In modern prokaryotes the donor DNA is transferred by conjugation, transduction or transformation. Transformation is hypothesized to be the ancestral mechanism, in which DNA from one prokaryote is released into the surrounding medium and then taken up by another prokaryotic cell. One theory on how meiosis arose is that it evolved from transformation. According to this view, the evolutionary transition from prokaryotic sex to eukaryotic sex was continuous.

Transformation, like meiosis, is a complex process requiring the function of numerous gene products. A key similarity between prokaryotic sex and eukaryotic sex is that DNA originating from two different individuals (parents) join up so that homologous sequences are aligned with each other, and this is followed by exchange of genetic information (a process called genetic recombination). After the new recombinant chromosome is formed it is passed on to progeny.

When genetic recombination occurs between DNA molecules originating from different parents, the recombination process is catalyzed in prokaryotes and eukaryotes by enzymes that have similar functions and that are evolutionarily related. One of the most important enzymes catalyzing this process in bacteria is referred to as RecA, and this enzyme has two functionally similar counterparts that act in eukaryotic meiosis, RAD51 and DMC1.

Support for the theory that meiosis arose from prokaryotic transformation comes from the increasing evidence that early diverging lineages of eukaryotes have the core genes for meiosis. This implies that the precursor to meiosis was already present in the prokaryotic ancestor of eukaryotes. For instance the common intestinal parasite *Giardia intestinalis*, a simple eukaryotic protozoan was, until recently, thought to be descended from an early diverging eukaryotic lineage that lacked sex. However, it has since been shown that *G. intestinalis* contains within its genome a core set of genes that function in meiosis, including five genes that function only in meiosis. In addition, *G. intestinalis* was recently found to undergo a specialized, sex-like process involving meiosis gene homologs. This evidence, and other similar examples, suggest that a primitive form of meiosis, was present in the common ancestor of all eukaryotes, an ancestor that arose from antecedent prokaryote.

Theory that Meiosis Evolved from Mitosis

Mitosis is the normal process in eukaryotes for cell division, duplicating chromosomes and segregating one of the two copies into each of the two daughter cells, and contrast with meiosis. The mitosis theory states that meiosis evolved from mitosis. On this theory, early eukaryotes evolved mitosis first, became established and only then did meiosis and sexual reproduction arise.

Support for the idea that meiosis arose from mitosis is the observation that some features, such as the meiotic spindles that draw chromosome sets into separate daughter cells upon cell division, and processes regulating cell division, employ the same, or similar, molecular machinery. Yet there is no compelling evidence for a period in the early evolution of eukaryotes during which meiosis and accompanying sexual capability did not yet exist.

In addition, as noted by Wilkins and Holliday, there are four novel steps needed in meiosis that are not present in mitosis. These are: (1) pairing of homologous chromosomes, (2) extensive recombination between homologs; (3) suppression of sister chromatid separation in the first meiotic division; and (4) avoiding chromosome replication during the second meiotic division. They note that the simultaneous appearance of these steps appears to be impossible, and the selective advantage for separate mutations to cause these steps is problematic, because the entire sequence is required for reliable production of a set of haploid chromosomes.

Sharing of Components During the Evolution of Meiosis and Mitosis

If meiosis arose from prokaryotic transformation, during the early evolution of eukaryotes, mitosis and meiosis could have evolved in parallel, with both processes using common molecular components, where mitosis evolved from the molecular machinery used by prokaryotes for DNA replication and segregation, and meiosis evolved from the prokaryotic sexual process of transformation, but meiosis also made use of the evolving molecular machinery for DNA replication and segregation.

Function

Stress Induces Sex in Protists and Bacteria

Abundant evidence indicates that facultative sexual eukaryotes tend to undergo sexual reproduction under stressful conditions. For instance, the budding yeast *Saccharomyces cerevisiae* (a single-celled fungus) reproduces mitotically (asexually) as diploid cells when nutrients are abundant, but switches to meiosis (sexual reproduction) under starvation conditions. The unicellular green alga, *Chlamydomonas reinhardtii* grows as vegetative cells in nutrient rich growth medium, but depletion of a source of nitrogen in the medium leads to gamete fusion, zygote formation and meiosis. The fission yeast *Schizosaccharomyces pombe*, treated with H2O2 to cause oxidative stress, substantially increases the proportion of cells which undergo meiosis. The simple multicellular eukaryote *Volvox carteri* undergoes sex in response to oxidative stress or stress from heat shock. These examples, and others, suggest that, in simple single-celled and multicellular eukaryotes, meiosis is an adaptation to respond to stress.

Prokaryotic sex also appears to be an adaptation to stress. For instance, transformation occurs near the end of logarithmic growth, when amino acids become limiting in *Bacillus subtilis*, or in *Haemophilus influenzae* when cells are grown to the end of logarithmic phase. In *Streptococcus mutans* and other streptococci, transformation is associated with high cell density and biofilm formation. In *Streptococcus pneumoniae*, transformation is induced by the DNA damaging agent mitomycin C. These, and other, examples indicate that prokaryotic sex, like meiosis in simple eukaryotes, is an adaptation to stressful conditions. This observation suggests that the natural selection pressures maintaining meiosis in eukaryotes are similar to the selective pressures maintaining prokaryotic sex. This similarity suggests continuity, rather than a gap, in the evolution of sex from prokaryotes to eukaryotes.

DNA Repair as the Adaptive Advantage

Stress is, however, a general concept. What is it specifically about stress that needs to be overcome by meiosis? And what is the specific benefit provided by meiosis that enhances survival under stressful conditions?

Again there are two contrasting theories. In one theory, meiosis is primarily an adaptation for repairing DNA damage. Environmental stresses often lead to oxidative stress within the cell, which is well known to cause DNA damage through the production of reactive forms of oxygen, known as reactive oxygen species (ROS). DNA damages, if not repaired, can kill a cell by blocking DNA replication, or transcription of essential genes.

When only one strand of the DNA is damaged, the lost information (nucleotide sequence) can ordinarily be recovered by repair processes that remove the damaged sequence and fill the resulting gap by copying from the opposite intact strand of the double helix. However, ROS also cause a type of damage that is difficult to repair, referred to as double-strand damage. One common example of double-strand damage is the double-strand break. In this case, genetic information (nucleotide sequence) is lost from both strands in the damaged region, and proper information can only be obtained from another intact chromosome homologous to the damage chromosome. The process that the cell uses to accurately accomplish this type of repair is called recombinational repair.

Meiosis is distinct from mitosis in that a central feature of meiosis is the alignment of homologous chromosomes followed by recombination between them. The two chromosomes which pair are referred to as non-sister chromosomes, since they did not arise simply from the replication of a parental chromosome. Recombination between non-sister chromosomes at meiosis is known to be a recombinational repair process that can repair double-strand breaks and other types of double-strand damage. In contrast, recombination between sister chromosomes cannot repair double-strand damages arising prior to the replication which produced them. Thus on this view, the adaptive advantage of meiosis is that it facilitates recombinational repair of DNA damage that is otherwise difficult to repair, and that occurs as a result of stress, particularly oxidative stress. If left unrepaired, this damage would likely be lethal to gametes and inhibit production of viable progeny.

Even in multicellular eukaryotes, such as humans, oxidative stress is a problem for cell survival. In this case, oxidative stress is a byproduct of oxidative cellular respiration occurring during metabolism in all cells. In humans, on average, about 50 DNA double-strand breaks occur per cell in each cell generation. Meiosis, which facilitates recombinational repair between non-sister chromosomes, can efficiently repair these prevalent damages in the DNA passed on to germ cells, and consequently prevent loss of fertility in humans. Thus with the theory that meiosis arose from prokaryotic sex, recombinational repair is the selective advantage of meiosis in both single celled eukaryotes and muticellular eukaryotes, such as humans.

Genetic Diversity of Offspring as the Adaptive Advantage

On the other view, stress is a signal to the cell that the environment is becoming adverse. Under this new condition, it may be beneficial to produce progeny that differ from the parent in their genetic make up. Among these varied progeny, some may be more adapted to the changed condition than their parents. Meiosis generates genetic variation in the diploid cell, in part by the exchange of genetic information between the pairs of chromosomes after they align (recombination). Thus, on this view, the advantage of meiosis is that it facilitates the generation of genomic diversity among progeny, allowing adaptation to adverse changes in the environment.

However, in the presence of a fairly stable environment, individuals surviving to reproductive age have genomes that function well in their current environment. They raise the question of why such individuals should risk shuffling their genes with those of another individual, as occurs during meiotic recombination? Considerations such as this have led many investigators to question whether genetic diversity is the adaptive advantage of sex.

References

- Michod, RE; Levin, BE, eds. (1987). The Evolution of sex: An examination of current ideas. Sunderland, Massachusetts: Sinauer Associates. ISBN 978-0878934584.
- Michod, RE (1994). Eros and Evolution: A Natural Philosophy of Sex. Perseus Books. ISBN 978-0201407549.4
- Harris Bernstein; Carol Bernstein & Richard E. Michod (2011). “Meiosis as an Evolutionary Adaptation for DNA Repair”. In Inna Kruman. DNA repair. InTech. doi:10.5772/1751. ISBN 978-953-307-697-3.

4

Errors and Variations of Mitosis

The errors and variations discussed in this chapter are aneuploidy, anaphase lag and endoreduplication. Aneuploidy is when a cell contains an incorrect number of chromosomes. An extra or a missing chromosome can cause a number of genetic disorders. The topics discussed in the chapter are of great importance to broaden the existing knowledge on mitosis.

Aneuploidy

Aneuploidy is the presence of an abnormal number of chromosomes in a cell, for example a human cell having 45 or 47 chromosomes instead of the usual 46. It does not include a difference of one or more complete sets of chromosomes, which is called *euploidy*. An extra or missing chromosome is a common cause of genetic disorders, including some human birth defects. Some cancer cells also have abnormal numbers of chromosomes. Aneuploidy originates during cell division when the chromosomes do not separate properly between the two cells.

Different species normally have different numbers of chromosomes from one another, and the terms "aneuploid" and "polyploid" refer to the chromosome number being different from the usual number for that species.

Chromosome abnormalities are detected in 1 of 160 live human births. Apart from sex chromosome disorders, most cases of aneuploidy result in death of the developing fetus (miscarriage); the most common extra autosomal chromosomes among live births are 21, 18 and 13.

Chromosomes

Most cells in the human body have 23 pairs of chromosomes, or a total of 46 chromosomes. (The sperm and egg, or gametes, each have 23 unpaired chromosomes, and red blood cells have no nucleus and no chromosomes.)

One copy of each pair is inherited from the mother and the other copy is inherited from the father. The first 22 pairs of chromosomes (called autosomes) are numbered from 1 to 22, from largest to smallest. The 23rd pair of chromosomes are the sex chromosomes. Normal females have two X chromosomes, while normal males have one X chromosome and one Y chromosome. The characteristics of the chromosomes in a cell as they are seen under a light microscope are called the karyotype.

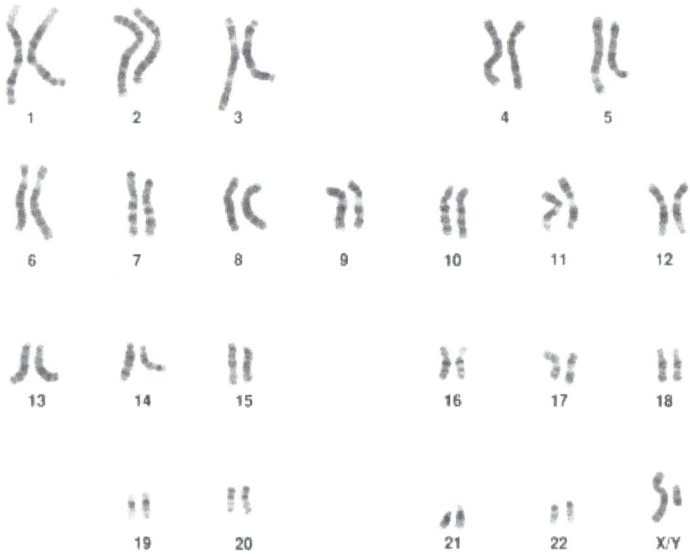

During meiosis, when germ cells divide to create sperm and egg (gametes), each half should have the same number of chromosomes. But sometimes, the whole pair of chromosomes will end up in one gamete, and the other gamete will not get that chromosome at all.

Most embryos cannot survive with a missing or extra autosome (numbered chromosome) and are spontaneously aborted. The most frequent aneuploidy in humans is trisomy 16, although fetuses affected with the full version of this chromosome abnormality do not survive to term (it is possible for surviving individuals to have the mosaic form, where trisomy 16 exists in some cells but not all). The most common aneuploidy that infants can survive with is trisomy 21, which is found in Down syndrome, affecting 1 in 800 births. Trisomy 18 (Edwards syndrome) affects 1 in 6,000 births, and trisomy 13 (Patau syndrome) affects 1 in 10,000 births. 10% of infants with trisomy 18 or 13 reach 1 year of age.

Changes in chromosome number may not necessarily be present in all cells in an individual. When aneuploidy is detected in a fraction of cells in an individual, it is called chromosomal mosaicism. In general, individuals who are mosaic for a chromosomal aneuploidy tend to have a less severe form of the syndrome compared to those with full trisomy. For many of the autosomal trisomies, only mosaic cases survive to term. However, mitotic aneuploidy may be more common than previously recognized in somatic tissues, and aneuploidy is a characteristic of many types of tumorigenesis.

Terminology

In the strict sense, a chromosome complement having a number of chromosomes other than 46 (in humans) is considered *heteroploid* while an exact multiple of the haploid chromosome complement is considered euploid.

Number of chromosomes	Name	Description
1	Monosomy	Monosomy refers to lack of one chromosome of the normal complement. Partial monosomy can occur in unbalanced translocations or deletions, in which only a portion of the chromosome is present in a single copy. Monosomy of the sex chromosomes (45,X) causes Turner syndrome.
2	Disomy	Disomy is the presence of two copies of a chromosome. For organisms such as humans that have two copies of each chromosome (those that are diploid), it is the normal condition. For organisms that normally have three or more copies of each chromosome (those that are triploid or above), disomy is an aneuploid chromosome complement. In uniparental disomy, both copies of a chromosome come from the same parent (with no contribution from the other parent).
3	Trisomy	Trisomy refers to the presence of three copies, instead of the normal two, of a particular chromosome. The presence of an extra chromosome 21, which is found in Down syndrome, is called trisomy 21. Trisomy 18 and Trisomy 13, known as Edwards Syndrome and Patau Syndrome, respectively, are the two other autosomal trisomies recognized in live-born humans. Trisomy of the sex chromosomes is possible, such as in (47,XXX), (47,XXY), and (47,XYY).
4/5	tetrasomy/pentasomy	Tetrasomy and pentasomy are the presence of four or five copies of a chromosome, respectively. Although rarely seen with autosomes, sex chromosome tetrasomy and pentasomy have been reported in humans, including XXXX, XXYY, XXXXX, XXXXY and XYYYY.

Mechanisms

Nondisjunction usually occurs as the result of a weakened mitotic checkpoint, as these checkpoints tend to arrest or delay cell division until all components of the cell are ready to enter the next phase. If a checkpoint is weakened, the cell may fail to 'notice' that a chromosome pair is not lined up on the mitotic plate, for example. In such a case, most chromosomes would separate normally (with one chromatid ending up in each cell), while others could fail to separate at all. This would generate a daughter cell lacking a copy and a daughter cell with an extra copy.

Completely inactive mitotic checkpoints may cause non-disjunction at multiple chromosomes, possibly all. Such a scenario could result in each daughter cell possessing a disjoint set of genetic material.

Merotelic attachment occurs when one kinetochore is attached to both mitotic spindle poles. One daughter cell would have a normal complement of chromosomes; the second would lack one. A third daughter cell may end up with the 'missing' chromosome.

Multipolar spindles: more than two spindle poles form. Such a mitotic division would result in one daughter cell for each spindle pole; each cell may possess an unpredictable complement of chromosomes.

Monopolar spindle: only a single spindle pole forms. This produces a single daughter cell with its copy number doubled.

A *tetraploid intermediate* may be produced as the end-result of the monopolar spindle mechanism. In such a case, the cell has double the copy number of a normal cell, and produces double the number of spindle poles as well. This results in four daughter cells with an unpredictable complement of chromosomes, but in the normal copy number.

Somatic Mosaicism in the Nervous System

Mosaicism for aneuploid chromosome content may be part of the constitutional make-up of the mammalian brain. In the normal human brain, brain samples from six individuals ranging from 2–86 years of age had mosaicism for chromosome 21 aneuploidy (average of 4% of neurons analyzed). This low-level aneuploidy appears to arise from chromosomal segregation defects during cell division in neuronal precursor cells, and neurons containing such aneuploid chromosome content reportedly integrate into normal circuits.

Somatic Mosaicism in Cancer

Aneuploidy is consistently observed in virtually all cancers. Somatic mosaicism occurs in virtually all cancer cells, including trisomy 12 in chronic lymphocytic leukemia (CLL) and trisomy 8 in acute myeloid leukemia (AML). However, these forms of mosaic aneuploidy occur through mechanisms distinct from those typically associated with genetic syndromes involving complete or mosaic aneuploidy, such as chromosomal instability (due to mitotic segregation defects in cancer cells). Therefore, the molecular processes that lead to aneuploidy are targets for the development of cancer drugs. Both resveratrol and aspirin have been found *in vivo* (in mice) to selectively destroy tetraploid cells that may be precursors of aneuploid cells, and activate AMPK, which may be involved in the process.

Alteration of normal mitotic checkpoints are also important tumorogenic events, and these may directly lead to aneuploidy. Loss of tumor suppressor p53 gene often results in genomic instability, which could lead to the aneuploidy genotype.

In addition, genetic syndromes in which an individual is predisposed to breakage of chromosomes (chromosome instability syndromes) are frequently associated with increased risk for various types of cancer, thus highlighting the role of somatic aneuploidy in carcinogenesis. It has been suggested that aneuploidy might directly contribute to carcinogenesis by disrupting the asymmetric division of adult stem cells, thereby leaving those cells capable of "limitless expansion".

Partial Aneuploidy

The terms "partial monosomy" and "partial trisomy" are used to describe an imbalance of genetic material caused by loss or gain of part of a chromosome. In particular, these terms would be used in the situation of an unbalanced translocation, where an individual carries a derivative chromosome formed through the breakage and fusion of two different chromosomes. In this situation, the individual would have three copies of part of one chromosome (two normal copies and the portion that exists on the derivative chromosome) and only one copy of part of the other chromosome involved in the derivative chromosome.

Aneuploidogens

Agents capable of causing aneuploidy are called aneuploidogens. Many mutagenic carcinogens are aneuploidogens. X-rays, for example, may cause aneuploidy by fragmenting the chromosome; it may also target the spindle apparatus. Other chemicals such as colchicine can also produce aneuploidy by affecting microtubule polymerization.

Exposure of males to lifestyle, environmental and/or occupational hazards may increase the risk of spermatozoa aneuploidy. Tobacco smoke contains chemicals that cause DNA damage. Smoking also can induce aneuploidy. For instance, smoking increases chromosome 13 disomy in spermatozoa by 3-fold, and YY disomy by 2-fold.

Occupational exposure to benzene is associated with a 2.8-fold increase of XX disomy and a 2.6-fold increase of YY disomy in spermatozoa.

Pesticides are released to the environment in large quantities so that most individuals have some degree of exposure. The insecticides fenvalerate and carbaryl have been reported to increase spermatozoa aneuploidy. Occupational exposure of pesticide factory workers to fenvalerate is associated with increased spermatozoa DNA damage. Exposure to fenvalerate raised sex chromosome disomy 1.9-fold and disomy of chromosome 18 by 2.6-fold. Exposure of male workers to carbaryl increased DNA fragmentation in spermatozoa, and also increased sex chromosome disomy by 1.7-fold and chromosome 18 disomy by 2.2-fold.

Humans are exposed to perfluorinated compounds (PFCs) in many commercial products. Men contaminated with PFCs in whole blood or seminal plasma have spermatozoa with increased levels of DNA fragmentation and chromosomal aneuploidies.

Diagnosis

Germline aneuploidy is typically detected through karyotyping, a process in which a sample of cells is fixed and stained to create the typical light and dark chromosomal banding pattern and a picture of the chromosomes is analyzed. Other techniques include Fluorescence In Situ Hybridization (FISH), quantitative PCR of Short Tandem

Repeats, quantitative fluorescence PCR (QF-PCR), quantitative PCR dosage analysis, Quantitative Mass Spectrometry of Single Nucleotide Polymorphisms, and Comparative Genomic Hybridization (CGH).

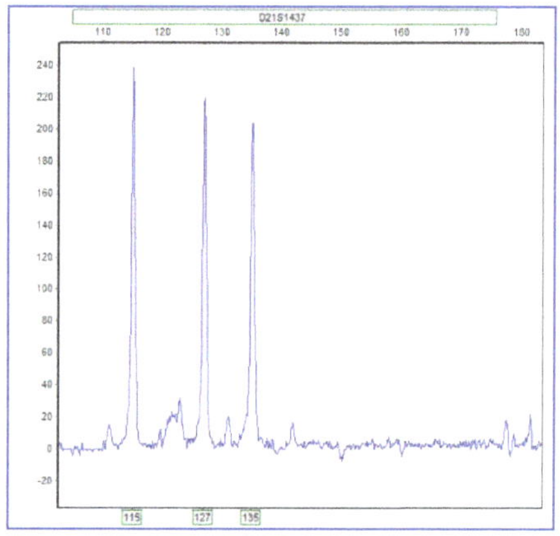

Example of Trisomy 21 detected via quantitative PCR Short Tandem Repeat assay

These tests can also be performed prenatally to detect aneuploidy in a pregnancy, through either amniocentesis or chorionic villus sampling. Pregnant women of 35 years or older are offered prenatal diagnosis because the chance of chromosomal aneuploidy increases as the mother's age increases.

Recent advances have allowed for less invasive testing methods based on the presence of fetal genetic material in maternal blood.

Types

key	
color	**significance**
	lethal
	normal male phenotype
	Klinefelter syndrome (abnormal male)
	polysomy X and/or Y, (abnormal male)
	normal female phenotype
	Turner's syndrome (abnormal female)
	tetrasomy X, pentasomy X, (abnormal female)

Non-autosomal						
	o	**X**	**XX**	**XXX**	**XXXX**	**XXXXX**
o	oo	Xo	XX	XXX	XXXX	XXXXX
Y	Yo	XY	XXY	XXXY	XXXXY	XXXXXY
YY	YY	XYY	XXYY	XXXYY	XXXXYY	XXXXXYY
YYY	YYY	XYYY	XXYYY	XXXYYY	XXXXYYY	XXXXXYYY

YYYY	YYYY	XYYYY	XXYYYY	XXXYYYY	XXXXYYYY	XXXXXYYYY
YYYYY	YYYYY	XYYYYY	XXYYYYY	XXXYYYYY	XXXXYYYYY	XXXXXYYYYY

key	
color	**significance**
orange	case where complete non-mosaic trisomy can never survive to term
yellow	case where complete non-mosaic trisomy can occasionally (barring other complications) survive to term
green	case where complete non-mosaic trisomy can always (barring other complications) survive to term

Autosomal		
#	**monosomy**	**trisomy**
1	1p36 deletion syndrome 1q21.1 deletion syndrome	Trisomy 1
2	2q37 deletion syndrome	Trisomy 2
3		Trisomy 3
4	Wolf-Hirschhorn syndrome	Trisomy 4
5	Cri du chat 5q deletion syndrome	Trisomy 5
6		Trisomy 6
7	Williams syndrome	Trisomy 7
8	Monosomy 8p Monosomy 8q	Trisomy 8
9	Alfi's syndrome Kleefstra syndrome	Trisomy 9
10	Monosomy 10p Monosomy 10q	Trisomy 10
11	Jacobsen syndrome	Trisomy 11
12		Trisomy 12
13		Patau syndrome
14		Trisomy 14
15	Angelman syndrome Prader–Willi syndrome	Trisomy 15
16		Trisomy 16
17	Miller-Dieker syndrome Smith-Magenis syndrome	Trisomy 17
18	Distal 18q- Proximal 18q-	Edwards syndrome
19		Trisomy 19

20		Trisomy 20
21		Down syndrome
22	DiGeorge syndrome Phelan-McDermid syndrome 22q11.2 distal deletion syndrome	Cat eye syndrome Trisomy 22

Anaphase Lag

Anaphase lag describes a delayed movement during anaphase, where one homologous chromosome in meiosis or one chromatid in mitosis fails to connect to the spindle apparatus, or is tardily drawn to its pole and fails to be included in the reforming nucleus. Instead, the chromosome forms a micronucleus in the cytoplasm and is lost from the cell. The lagging chromosome is not incorporated into the nucleus of one of the daughter cells, resulting in one normal daughter cell and one with monosomy. Anaphase lag is one of several causes of aneuploidy and one of several causes of mosaicism. Anaphase lag can also cause a rescue of the daughter cell if the cell was originally trisomy.

Endoreduplication

Endoreplication (also referred to as endoreduplication or polytenization) is replication of the nuclear genome in the absence of cell division, which leads to elevated nuclear gene content and polyploidy. Endoreplication can be understood simply as a variant form of the mitotic cell cycle (G1-S-G2-M) in which mitosis is aborted prior to cytokinesis or circumvented entirely, due in part to modulation of cyclin-dependent kinase (CDK) activity. Examples of endoreplication characterized in arthropod, mammalian, and plant species suggest that it is a universal developmental mechanism responsible for the differentiation and morphogenesis of cell types that fulfill an array of biological functions. While endoreplication is often limited to specific cell types in animals, it is considerably more widespread in plants, such that polyploidy can be detected in the majority of plant tissues.

Examples in Nature

Endoreplicating cell types that have been studied extensively in model organisms

Organism	Cell Type	Biological Function
fly	larval tissues (incl. salivary glands)	secretion, embryogenesis
fly	ovarian follicle, nurse cells	nourishment, protection of oocytes

rodent	megakaryocyte	platelet formation
rodent	trophoblast giant cell	placental development, nourishment of embryo
plant	trichome	defense from herbivory, homeostasis
plant	leaf epidermal cell	leaf size, structure
plant	endosperm	nourishment of embryo
nematode	hypodermis	secretion, body size

Endocycling Vs. Endomitosis

Endoreplication typically results in elevated cellular DNA content, but the manner in which the genetic material is configured varies depending on whether mitotic events are allowed to occur. Endocycling is a form of endoreplication in which the cell largely avoids mitosis and duplicated chromatids remain physically associated. Repeated rounds of endocycling can lead to the production of polytene chromosomes in which sister chromatids are tightly configured in parallel arrays. In contrast, endomitosis is a form of endoreplication in which cells undergo aspects of mitosis but fail to execute telophase and/or cytokinesis. Duplicated chromosomes produced by endomitosis exist as discrete units in a single polyploid nucleus or may be packaged into separate nuclei, depending on the phase at which mitosis is aborted. In certain instances (e.g. trophoblast giant cells), the distinction between endocycling and endomitosis is clouded, as the cell undergoes early mitotic events such as chromosome condensation that tend to diminish the physical association among homologous chromosomes. It should also be noted that endocycling does not always involve replication of the entire genome, and that certain regions are often replicated more frequently than others.

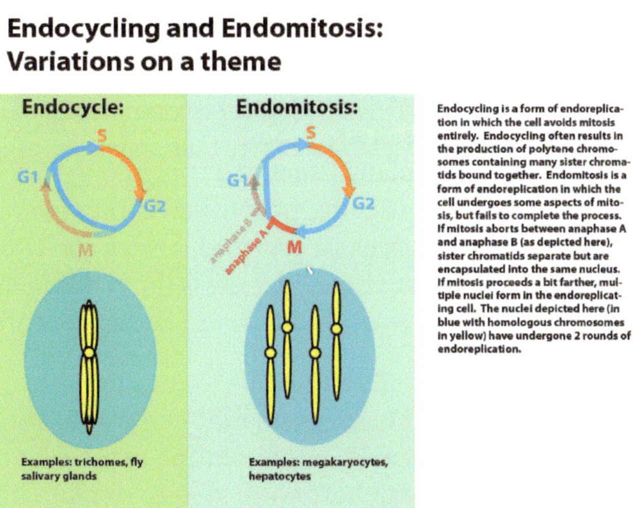

Endocycling Vs. Endomitosis

Biological Significance

Based on the wide array of cell types in which endoreplication occurs, a variety of hypotheses have been generated to explain the functional importance of this phenomenon. Unfortunately, experimental evidence to support these conclusions is somewhat limited:

Cell/Organism Size: Cell ploidy often correlates with cell size, and in some instances, disruption of endoreplication results in diminished cell and tissue size suggesting that endoreplication may serve as a mechanism for tissue growth. Relative to mitosis, endoreplication does not require cytoskeletal rearrangement or the production of new cell membrane and it often occurs in cells that have already differentiated. As such it may represent an energetically efficient alternative to cell proliferation among differentiated cell types that can no longer afford to undergo mitosis. While evidence establishing a connection between ploidy and tissue size is prevalent in the literature, contrary examples also exist.

Cell Differentiation: In developing plant tissues the transition from mitosis to endoreplication often coincides with cell differentiation and morphogenesis. However it remains to be determined whether endoreplication and polypoidy contribute to cell differentiation or vice versa. Interestingly, targeted inhibition of endoreplication in trichome progenitors results in the production of multicellular trichomes that exhibit relatively normal morphology, but ultimately dedifferentiate and undergo absorption into the leaf epidermis. This result suggests that endoreplication and polyploidy may be required for the maintenance of cell identity.

Oogenesis and Embryonic Development: Endoreplication is commonly observed in cells responsible for the nourishment and protection of oocytes and embryos. It has been suggested that increased gene copy number might allow for the mass production of proteins required to meet the metabolic demands of embryogenesis and early development. Consistent with this notion, mutation of the Myc oncogene in *Drosophila* follicle cells results in reduced endoreplication and abortive oogenesis. However, reduction of endoreplication in maize endosperm has limited effect on the accumulation of starch and storage proteins, suggesting that the nutritional requirements of the developing embryo may involve the nucleotides that comprise the polyploid genome rather than the proteins it encodes.

Buffering the Genome: Another intriguing hypothesis is that endoreplication buffers against DNA damage and mutation because it provides extra copies of important genes. However, this notion is purely speculative and there is limited evidence to the contrary. For example, analysis of polyploid yeast strains suggests that they are more sensitive to radiation than diploid strains.

Stress Response: Research in plants suggests that endoreplication may also play a role in modulating stress responses. By manipulating expression of E2fe (a repressor of endocycling in plants), researchers were able to demonstrate that increased cell ploidy lessens the negative impact of drought stress on leaf size. Given that the sessile lifestyle of plants necessitates a capacity to adapt to environmental conditions, it is appealing to speculate that widespread polyploidization contributes to their developmental plasticity

Genetic Control of Endoreplication

The best-studied example of a mitosis-to-endocycle transition occurs in *Drosophila* follicle cells and is activated by Notch signaling. Entry into endocycles involves modulation of mitotic and S-phase cyclin-dependent kinase (CDK) activity. Inhibition of M-phase CDK activity is accomplished via transcriptional activation of Cdh/fzr and repression of the G2-M regulator string/cdc25. Cdh/fzr is responsible for activation of the anaphase-promoting complex (APC) and subsequent proteolysis of the mitotic cyclins. String/cdc25 is a phosphatase that stimulates mitotic cyclin-CDK complex activity. Upregulation of S-phase CDK activity is accomplished via transcriptional repression of the inhibitory kinase dacapo. Together, these changes allow for the circumvention of mitotic entry, progression through G1, and entry into S-phase. The induction of endomitosis in mammalian megakaryocytes involves activation of the c-mpl receptor by the thrombopoietin (TPO) cytokine and is mediated by ERK1/2 signaling. As with Drosophila follicle cells, endoreplication in megakaryocytes results from activation of S-phase cyclin-CDK complexes and inhibition of mitotic cyclin-CDK activity.

Genetic model for Notch regulation of the mitosis-to-endocycle transition in Drosophila egg follicles:

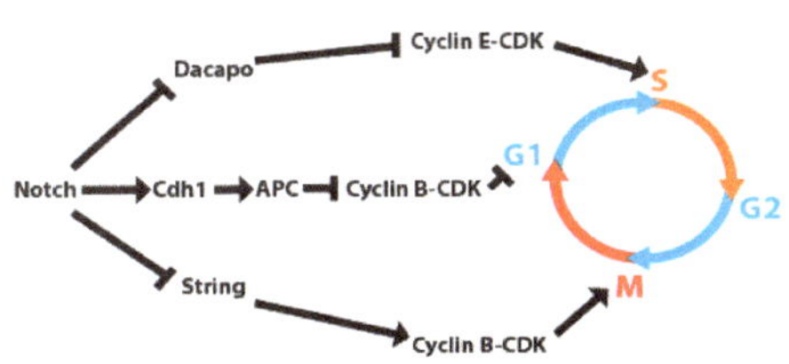

Transcriptional and post-transcriptional forms of regulation are included. Arrows indicate positive regulation, T-bars indicate negative regulation. Not all relationships are direct.

Notch Regulation of Endocycling

Entry into S-phase during endoreplication (and mitosis) is regulated through the formation of a prereplicative complex (pre-RC) at replication origins, followed by recruitment and activation of the DNA replication machinery. In the context of endoreplication these events are facilitated by an oscillation in cyclin E-Cdk2 activity. Cyclin E-Cdk2 activity drives the recruitment and activation of the replication machinery, but it also inhibits pre-RC formation, presumably to ensure that only one round of replication occurs per cycle. Failure to maintain control over pre-RC formation at replication origins results in a phenomenon known as "rereplication" which is common in cancer cells. The mechanism by which cyclin E-Cdk2 inhibits pre-RC formation involves downregulation of APC-Cdh1-mediated proteolysis and accumulation of the protein Geminin, which is responsible for sequestration of the pre-RC component Cdt1.

Oscillations in Cyclin E-Cdk2 activity are modulated via transcriptional and post-transcriptional mechanisms. Expression of cyclin E is activated by E2F transcription factors that were shown to be required for endoreplication. Recent work suggests that observed oscillations in E2F and cyclin E protein levels result from a negative-feedback loop involving Cul4-dependent ubiquitination and degradation of E2F. Post-transcriptional regulation of cyclin E-Cdk2 activity involves Ago/Fbw7-mediated proteolytic degradation of cyclin E and direct inhibition by factors such as Dacapo and p57. True endomitosis in the anther tapetum of the liliaceous plant Eremurus is described. The nuclear membrane does not disappear, but during metaphase the chromosomes are condensed, often considerably more than in normal mitosis. When the pollen mother cells (PMCs) go through the last premeiotic mitosis, the tapetal cells have one diploid nucleus which divides while the cell remains undivided. The two diploid nuclei may undergo an endomitosis and the resulting tetraploid nuclei a second endomitosis. An alternative pathway is an ordinary mitosis-again without cell division instead of one of the endomitotic cycles. The cytological picture in the tapetum is further complicated by restitution in anaphase and fusion of metaphase and anaphase groups during mitosis, processes which could give rise to cells with one, two, or three nuclei, instead of the expected two or four. No sign of the so-called "inhibited" mitosis is seen in these tapetal cells. When the PMCs are in leptotene-zygotene, very few tapetal nuclei are in endomitosis. When the PMCs have reached diplotene, almost 100% of cells which are not in interphase show an endomitotic stage.

Endoreplication and Oncogenesis

Polyploidy and aneuploidy are common phenomena in cancer cells. Given that oncogenesis and endoreplication likely involve subversion of common cell cycle regulatory mechanisms, a thorough understanding of endoreplication may provide important insights for cancer biology.

Premeiotic Endomitosis in Unisexual Vertebrates

The unisexual salamanders (genus *Ambystoma*) are the oldest known unisexual vertebrate lineage, having arisen about 5 million years ago. In these polyploid unisexual females, an extra premeiotic endomitotic replication of the genome, doubles the number of chromosomes. As a result, the mature eggs that are produced subsequent to the two meiotic divisions have the same ploidy as the somatic cells of the adult female salamander. Synapsis and recombination during meiotic prophase I in these unisexual females is thought to ordinarily occur between identical sister chromosomes and occasionally between homologous chromosomes. Thus little, if any, genetic variation is produced. Recombination between homeologous chromosomes occurs rarely, if at all. Since production of genetic variation is weak, at best, it is unlikely to provide a benefit sufficient to account for the maintenance of meiosis for millions of years. Perhaps the efficient recombinational repair of DNA damages at each generation provided by meiosis has been a sufficient advantage to maintain meiosis.

Consequences of Errors in Mitosis

Binucleated cells

Binucleated cells are cells that contain two nuclei. This type of cell is most commonly found in cancer cells and may arise from a variety of causes. Binucleation can be easily visualized through staining and microscopy. In general, binucleation has negative effects on cell viability and subsequent mitosis.

They also occur physiologically in hepatocytes, chondrocytes and in fungi (dikaryon).

Causes

- Cleavage furrow regression: Cells divide and almost complete division but then the cleavage furrow begins to regress and the cells merge. This is thought to be caused by nondisjunction in chromosomes but the mechanism by which it occurs is not well understood.

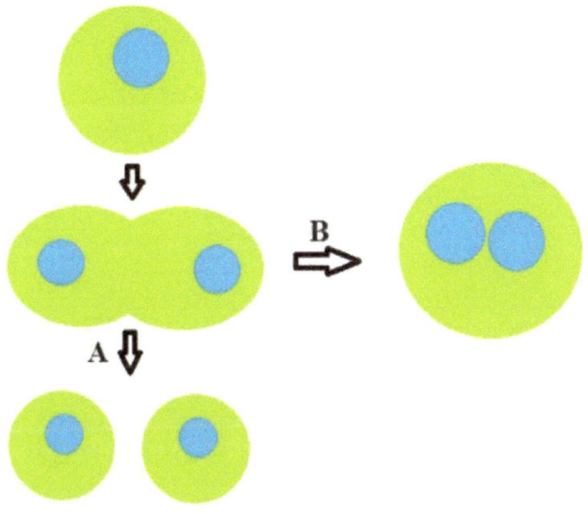

Cartoon of cleavage furrow formation and resolution. (A) depicts a normal cleavage furrow resolution. (B) depicts a cleavage furrow regression leading to a binucleated cell

- Failed cytokinesis: The cell can fail to form a cleavage furrow, leading to both nuclei remaining in one cell.
- Multipolar spindles: Cells contain three or more centrioles, resulting in multiple poles. This leads to the cells pulling chromosomes in many directions that end in multiple nuclei found in one cell.
- Merging of newly formed cells: Two cells that have just finished cytokinesis merge into one another. This process is not entirely understood.

Medical Relevance

Detection

Binucleated cells can be observed using microscopy. Cells must first be fixed to arrest them wherever they are in the cell cycle and to keep their structures from degrading. Their nuclei and tubulin must next be made visible so that binucleation can be identified. DAPI is a dye that binds to DNA and fluoresces blue. For this reason, it is particularly useful at labeling nuclei. Antibody probes can be used to label tubulin fluorescently. The immunofluorescence may then be observed with microscopy. Binucleated cells are most easily identified by viewing tubulin, which surrounds the two nuclei in the cell. Binucleated cells may be mistaken for two cells in close proximity when viewing only nuclei.

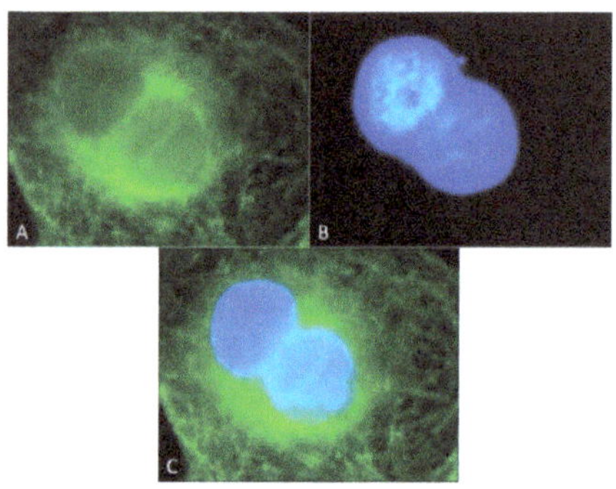

The binucleated cell above is an oral squamous cell carcinoma, cell line UPCI-SCC-103. A: The cancer cell treated with antibodies against tubulin. B: The same cancer cell stained with DAPI, highlighting the nuclei. C: Both photos superimposed.

Cancer

Binucleation occurs at a much higher rate in cancer cells. Other identifying features of cancer cells include multipolar spindles, micronuclei, and chromatin bridge. However, the increased rate of binucleation is usually not high enough to make it a conclusive diagnostic tool.

Effects

The fate of binucleated cells depends largely on the type of cell they originated from. A large percentage of binucleated cells arising from normal cells remain in interphase and never enter mitosis again. Cells that contain many mutations before they become binucleate are much more likely to proceed through subsequent rounds of mitosis. One study found that more than 50% of binucleated cells never entered mitosis again while greater than 95% of cancer cells were able to proceed through mitosis. Subsequent rounds of mitosis in binucleated cells have much higher rates of errors in chromosomal disjunction making it much more likely for cells to accumulate mutations.

Polytene Chromosome

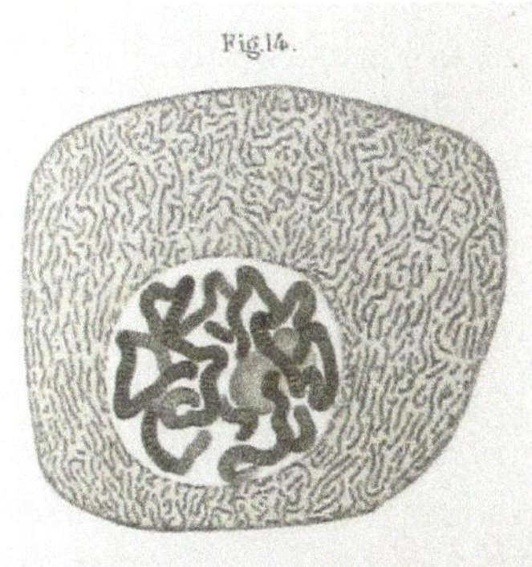

Polytene chromosomes in a *Chironomus* salivary gland cell

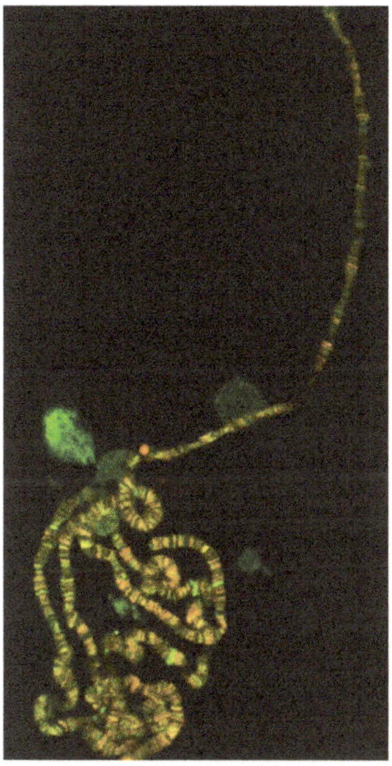

Polytene chromosome

Polytene chromosomes are oversized chromosomes which have developed from standard chromosomes and are commonly found in the salivary glands of *Drosophila melanogaster*. Specialized cells undergo repeated rounds of DNA replication without cell division (endomitosis), to increase cell volume, forming a giant polytene chromosome. Polytene chromosomes form when multiple rounds of replication produce many sister chromatids that remain fused together.

Function

In addition to increasing the volume of the cells' nuclei and causing cell expansion, polytene cells may also have a metabolic advantage as multiple copies of genes permits a high level of gene expression. In *Drosophila melanogaster*, for example, the chromosomes of the larval salivary glands undergo many rounds of endoreduplication, to produce large amounts of glue before pupation. Another example within the organism itself is the tandem duplication of various polytene bands located near the centromere of the X chromosome which results in the Bar phenotype of kidney-shaped eyes.

Structure

Polytene chromosomes have characteristic light and dark banding patterns that can be used to identify chromosomal rearrangements and deletions. Dark banding frequently corresponds to inactive chromatin, whereas light banding is usually found at areas with higher transcriptional activity. The banding patterns of the polytene chromosomes of *Drosophila melanogaster* were sketched in 1935 by Calvin B. Bridges, in such detail that his maps are still widely used today. The banding patterns of the chromosomes are especially helpful in research, as they provide an excellent visualization of transcriptionally active chromatin and general chromatin structure. For example, the polytene chromosomes in *Drosophila* have been used to support the theory of genomic equivalence, which states that all of the cells in the body maintain the same genome. Chromosome puffs are diffused uncoiled regions of the polytene chromosome that are sites of RNA transcription. A Balbiani ring is a large chromosome puff. Polytene chromosomes are about 200 μm in length. The chromonema of these chromosomes divide but do not separate. Therefore, they remain together to become large in size. Another form of chromosomal enlargement that provides for increased transcription is the lampbrush chromosome.

History

Polytene chromosomes were originally observed in the larval salivary glands of *Chironomus* midges by Balbiani in 1881, but the hereditary nature of these structures was not confirmed until they were studied in *Drosophila melanogaster* in the early 1930s by Emil Heitz and Hans Bauer. They are known to occur in secretory tissues of other dipteran insects such as the Malpighian tubules of *Sciara* and also in protists, plants, mammals, or in cells from other insects. Some of the largest polytene chromosomes described thus far occur in larval salivary gland cells of the Chironomid genus *Axarus*.

Polytene chromosomes are also used to identify the species of Chironomid larvae that are notoriously difficult to identify. Each morphologically distinct group of larvae consists of a number of morphologically identical (sibling) species that can only be identified by rearing adult males or by cytogenetic analysis of the polytene chromosomes of the larvae. Karyotypes are used to confirm the presence of specific species and to study genetic diversity in species with a wide range.

References

- Griffiths, Anthony JF; Miller, Jeffrey H; Suzuki, David T; Lewontin, Richard C; Gelbart, William

M (2000). "Chromosome Mutation II: Changes in Chromosome Number". An Introduction to Genetic Analysis (7th ed.). New York: W. H. Freeman. ISBN 978-0-7167-3520-5. Retrieved 2009-06-21.

- Bernstein C, Prasad AR, Nfonsam V, Bernstein H. (2013). DNA Damage, DNA Repair and Cancer, New Research Directions in DNA Repair, Prof. Clark Chen (Ed.), ISBN 978-953-51-1114-6.
- Hartwell, Leland; Leroy Hood; Michael L. Goldberg; Ann E. Reynolds; Lee M. Silver (2011). Genetics:From Genes to Genomes; Fourth Edition. New York, NY: McGraw-Hill. ISBN 978-0-07-352526-6.
- Templado C, Uroz L, Estop A (2013). "New insights on the origin and relevance of aneuploidy in human spermatozoa". Mol. Hum. Reprod. 19 (10): 634–43. doi:10.1093/molehr/gat039. PMID 23720770.

5

Cell Division and Cycle: A Comprehensive Study

The series of events that take place when a cell gets divided is known as cell division. The cell cycle is also divided into three periods: interphase, the mitotic phase, and cytokinesis. During interphase, the cell grows and accumulates nutrients, preparing itself for cell division, and then during the mitotic phase, the chromosomes separate leading the cell to its final stage, the cytokinesis. This section provides a study of cell division and cell cycle.

Cell Division Orientation

Cell division orientation is the direction along which the new daughter cells are formed. Cell division orientation is important for morphogenesis, cell fate and tissue homeostasis. Abnormalities in the cell division orientation leads to the malformations during development and cancerous tissues. Factors influenced the cell division orientation are cell shape, anisotropic localization of specific proteins and mechanical tensions.

Implication for Morphogenesis

Cell division orientation is one of the mechanisms that shapes tissue during development and morphogenesis. Along with cell shape changes, cell rearrangements, apoptosis and growth, oriented cell division modifies the geometry and topology of live tissue in order to create new organs and shape the organisms. Reproducible patterns of oriented cell divisions were described during mophogenesis of *Drosophila* embryos, *Drosophila* pupa, zebrafish embryos and mouse early embryos. Oriented cell divisions contribute to the tissue elongation and the release of mechanical stress. While in the first case oriented cell division akts as active contributor to the morphogenesis, the latter case is a passive response to the external mechanical tensions.

Implication for Tissue Homeostasis

In several tissues, such as columnar epithelium, the cells divide along the plane of the epithelium. Such divisions insert new formed cells in the epithelium layer. The disregulation of the orientation of cell divisions result in the creation of the cell out of epithelium and is observed at the initial stages of cancer.

Regulation

More than a century ago Oskar Hertwig proposed that the cell division orientation is determined by the shape of the cell (1884), known as Hertwig rule. In the epithelium the cells 'reads' its shape through the specific cell junction called tricellular junctions (TCJ). TCJ provide mechanical and geometrical clues for the spindle apparatus to ensure that cell divide along its long axis. Several factors could regulate cell shape and therefore orientation of cell division. Among these factors is the anisotropic mechanical stress. This stress could be the result of the external mechanical deformation of generated intracellularly by non-isotropic localization of specific proteins.

Cell Division

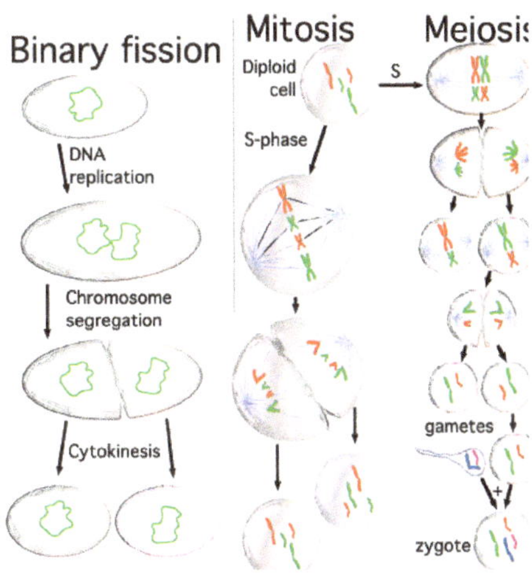

Three types of cell division

Cell division is the process by which a *parent cell* divides into two or more *daughter cells*. Cell division usually occurs as part of a larger cell cycle. In eukaryotes, there are two distinct types of cell division: a vegetative division, whereby each daughter cell is genetically identical to the parent cell (mitosis), and a reproductive cell division, whereby the number of chromosomes in the daughter cells is reduced by half, to produce haploid gametes (meiosis). Meiosis results in four haploid daughter cells by undergoing one round of DNA replication followed by two divisions: homologous chromosomes are separated in the first division, and sister chromatids are separated in the second division. Both of these cell division cycles are used in sexually reproducing organisms at some point in their life cycle, and both are believed to be present in the last eukaryotic common ancestor. Prokaryotes also undergo a vegetative cell division known as binary fission, where their genetic material is segregated equally into two daughter cells. All cell divisions, regardless of organism, are preceded by a single round of DNA replication.

For simple unicellular organisms such as the amoeba, one cell division is equivalent to reproduction – an entire new organism is created. On a larger scale, mitotic cell division can create progeny from multicellular organisms, such as plants that grow from cuttings. Cell division also enables sexually reproducing organisms to develop from the one-celled zygote, which itself was produced by cell division from gametes. And after growth, cell division allows for continual construction and repair of the organism. A human being's body experiences about 10 quadrillion cell divisions in a lifetime.

The primary concern of cell division is the maintenance of the original cell's genome. Before division can occur, the genomic information that is stored in chromosomes must be replicated, and the duplicated genome must be separated cleanly between cells. A great deal of cellular infrastructure is involved in keeping genomic information consistent between "generations".

Variants

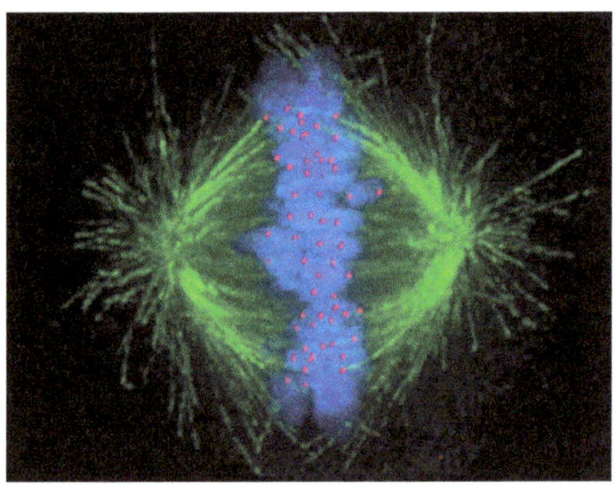

Image of the mitotic spindle in a human cell showing microtubules in green, chromosomes (DNA) in blue, and kinetochores in red.

Cells are classified into two main categories: simple, non-nucleated prokaryotic cells, and complex, nucleated eukaryotic cells. Owing to their structural differences, eukaryotic and prokaryotic cells do not divide in the same way. Also, the pattern of cell division that transforms eukaryotic stem cells into gametes (sperm cells in males or ova – egg cells – in females) is different from that of the somatic cell division in the cells of the body.

Cell division over 42 hours. The cells were directly imaged in the cell culture vessel, using non-invasive quantitative phase contrast time-lapse microscopy.

Degradation

Multicellular organisms replace worn-out cells through cell division. In some animals, however, cell division eventually halts. In humans this occurs on average, after 52 divisions, known as the Hayflick limit. The cell is then referred to as senescent. Cells stop dividing because the telomeres, protective bits of DNA on the end of a chromosome required for replication, shorten with each copy, eventually being consumed. Cancer cells, on the other hand, are not thought to degrade in this way, if at all. An enzyme called telomerase, present in large quantities in cancerous cells, rebuilds the telomeres, allowing division to continue indefinitely.

Cell Cycle

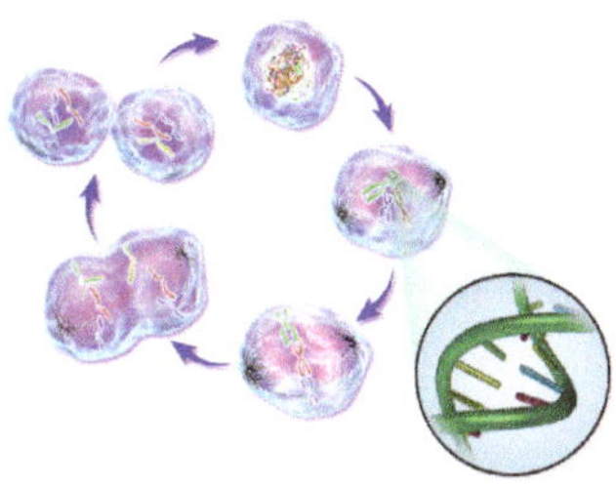

Life Cycle of the cell

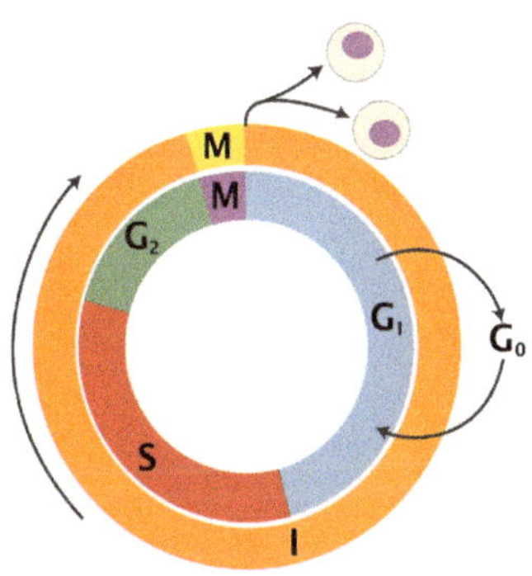

Schematic of the cell cycle. outer ring: I = Interphase, M = Mitosis; inner ring: M = Mitosis, G_1 = Gap 1, G_2 = Gap 2, S = Synthesis; not in ring: G_0 = Gap 0/Resting.

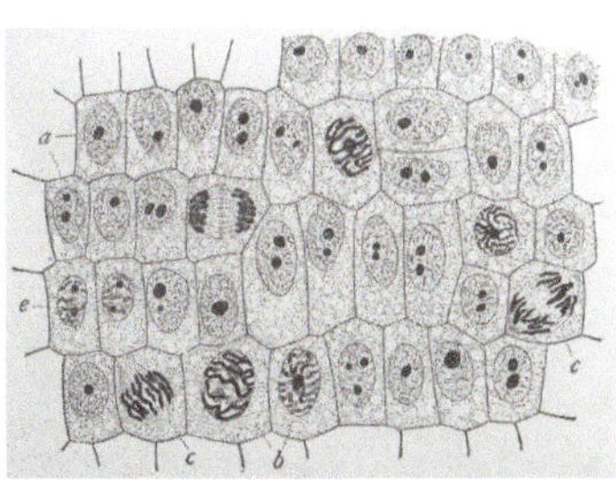

Onion (Allium) cells in different phases of the cell cycle. Growth in an organism is carefully controlled by regulating the cell cycle.

The cell cycle or cell-division cycle is the series of events that take place in a cell leading to its division and duplication of its DNA (DNA replication) to produce two daughter cells. In bacteria, which lack a cell nucleus, the cell cycle is divided into the B, C, and D periods. The B period extends from the end of cell division to the beginning of DNA replication. DNA replication occurs during the C period. The D period refers to the stage between the end of DNA replication and the splitting of the bacterial cell into two daughter cells. In cells with a nucleus, as in eukaryotes, the cell cycle is also divided into three periods: interphase, the mitotic (M) phase, and cytokinesis. During interphase, the cell grows, accumulating nutrients needed for mitosis, preparing it for cell division and duplicating its DNA. During the mitotic phase, the chromosomes separate. During the final stage, cytokinesis, the chromosomes and cytoplasm separate into two new daughter cells. To ensure the proper division of the cell, there are control mechanisms known as cell cycle checkpoints.

The cell-division cycle is a vital process by which a single-celled fertilized egg develops into a mature organism, as well as the process by which hair, skin, blood cells, and some internal organs are renewed. After cell division, each of the daughter cells begin the interphase of a new cycle. Although the various stages of interphase are not usually morphologically distinguishable, each phase of the cell cycle has a distinct set of specialized biochemical processes that prepare the cell for initiation of cell division.

Cell Cycle Phases

State	Description	Abbreviation	:	
Quiescent senescent/ resting phase	Gap 0	G_0	A resting phase where the cell has left the cycle and has stopped dividing.Cell cycle starts with this phase.	
	Interphase	Gap 1	G_1	Cells increase in size in Gap 1. The G_1 checkpoint control mechanism ensures that everything is ready for DNA synthesis.
		Synthesis	S	DNA replication occurs during this phase.
Gap 2		G_2	During the gap between DNA synthesis and mitosis, the cell will continue to grow. The G_2 checkpoint control mechanism ensures that everything is ready to enter the M (mitosis) phase and divide.	
Cell division	Mitosis	M	Cell growth stops at this stage and cellular energy is focused on the orderly division into two daughter cells. A checkpoint in the middle of mitosis (Metaphase Checkpoint) ensures that the cell is ready to complete cell division.	

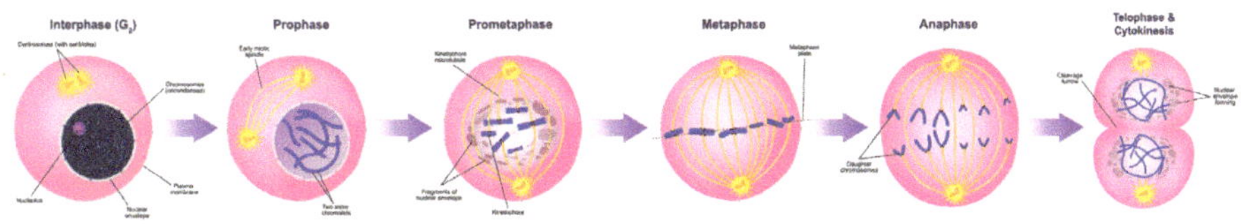

G_0 Phase (Quiescence)

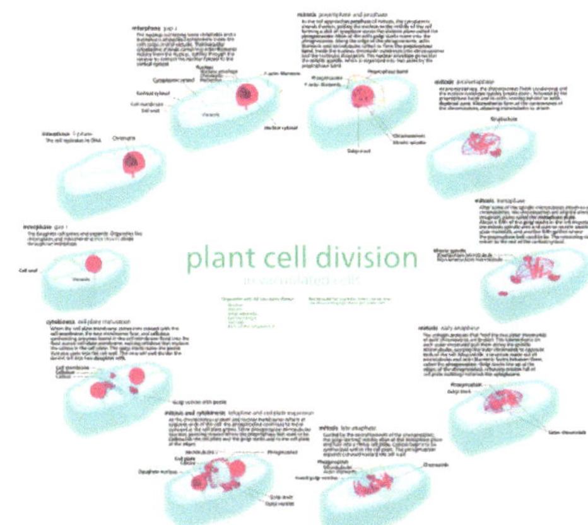

Plant cell cycle

The word "post-mitotic" is sometimes used to refer to both quiescent and senescent cells. Nonproliferative cells in multicellular eukaryotes generally enter the quiescent G_0 state from G_1 and may remain quiescent for long periods of time, possibly indefinitely (as is often the case for neurons). This is very common for cells that are fully differentiated. Cellular senescence occurs in response to DNA damage or degradation that would make a cell's progeny nonviable; for example, become cancerous. Some cells enter the G_0 phase semi-permanently e.g., some liver, kidney, stomach cells. Many cells do not enter G_0 and continue to divide throughout an organism's life, e.g. epithelial cells.

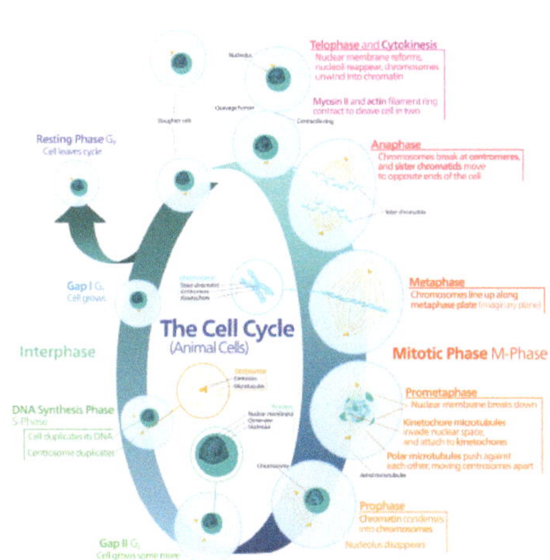

Animal cell cycle

Interphase (Intermitosis)

Before a cell can enter cell division, it needs to take in nutrients. All of the preparations are done during interphase. Interphase is a series of changes that takes place in a newly formed cell and its nucleus, before it becomes capable of division again. It is also called preparatory phase or intermitosis. Previously it was called resting stage because there is no apparent activity related to cell division.Typically interphase lasts for at least 90% of the total time required for the cell cycle.

Interphase proceeds in three stages, G_1, S, and G_2, followed by the cycle of mitosis and cytokinesis. The cell's nuclear chromosomes are duplicated during S phase.

G_1 Phase (Growth)

The first phase within interphase, from the end of the previous M phase until the beginning of DNA synthesis, is called G_1 (G indicating *gap*). It is also called the growth phase. During this phase, the biosynthetic activities of the cell, which are considerably slowed down during M phase, resume at a high rate. The duration of G_1 is highly variable, even among different cells of the same species. In this phase, the cell increases its supply of proteins, increases the number of organelles (such as mitochondria, ribosomes), and grows in size.

S Phase (DNA Replication)

The ensuing S phase starts when DNA replication commences; when it is completed, all of the chromosomes have been replicated, i.e., each chromosome has two (sister) chromatids. Thus, during this phase, the amount of DNA in the cell has effectively doubled, though the ploidy of the cell remains the same. During this phase, synthesis is completed as quickly as possible due to the exposed base pairs being sensitive to harmful external factors such as mutagens

G_2 Phase (Growth)

G2 phase is a period of protein synthesis and rapid cell growth to prepare the cell for mitosis.

Mitotic Phase (Chromosome Separation)

The relatively brief *M phase* consists of nuclear division (karyokinesis). It is a relatively short period of the cell cycle. M phase is complex and highly regulated. The sequence of events is divided into phases, corresponding to the completion of one set of activities and the start of the next. These phases are sequentially known as:

- prophase,
- metaphase,

- anaphase,
- telophase

Mitosis is the process by which a eukaryotic cell separates the chromosomes in its cell nucleus into two identical sets in two nuclei. During the process of mitosis the pairs of chromosomes condense and attach to fibers that pull the sister chromatids to opposite sides of the cell.

Mitosis occurs exclusively in eukaryotic cells, but occurs in different ways in different species. For example, animals undergo an "open" mitosis, where the nuclear envelope breaks down before the chromosomes separate, while fungi such as *Aspergillus nidulans* and *Saccharomyces cerevisiae* (yeast) undergo a "closed" mitosis, where chromosomes divide within an intact cell nucleus. Prokaryotic cells, which lack a nucleus, divide by a process called binary fission.

Cytokinesis Phase (Separation of All Cell Components)

Mitosis is immediately followed by cytokinesis, which divides the nuclei, cytoplasm, organelles and cell membrane into two cells containing roughly equal shares of these cellular components. Mitosis and cytokinesis together define the division of the mother cell into two daughter cells, genetically identical to each other and to their parent cell. This accounts for approximately 10% of the cell cycle.

Because cytokinesis usually occurs in conjunction with mitosis, "mitosis" is often used interchangeably with "M phase". However, there are many cells where mitosis and cytokinesis occur separately, forming single cells with multiple nuclei in a process called endoreplication. This occurs most notably among the fungi and slime moulds, but is found in various groups. Even in animals, cytokinesis and mitosis may occur independently, for instance during certain stages of fruit fly embryonic development. Errors in mitosis can either kill a cell through apoptosis or cause mutations that may lead to cancer.

Regulation of Eukaryotic Cell Cycle

Regulation of the cell cycle involves processes crucial to the survival of a cell, including the detection and repair of genetic damage as well as the prevention of uncontrolled cell division. The molecular events that control the cell cycle are ordered and directional; that is, each process occurs in a sequential fashion and it is impossible to "reverse" the cycle.

Role of Cyclins and CDKs

Two key classes of regulatory molecules, cyclins and cyclin-dependent kinases (CDKs), determine a cell's progress through the cell cycle. Leland H. Hartwell, R. Timothy Hunt, and Paul M. Nurse won the 2001 Nobel Prize in Physiology or Medicine for their discovery of these central molecules. Many of the genes encoding cyclins and CDKs are conserved among all eukaryotes, but in general more complex organisms have more elaborate cell cycle control systems that incorporate more individual components. Many of the relevant genes were first identified by studying yeast, especially *Saccharomyces cerevisiae*; genetic nomenclature in yeast dubs many of these genes *cdc* (for "cell division cycle") followed by an identifying number, e.g. *cdc25* or *cdc20*.

Nobel Laureate Paul Nurse

Cyclins form the regulatory subunits and CDKs the catalytic subunits of an activated heterodimer; cyclins have no catalytic activity and CDKs are inactive in the absence of a partner cyclin. When activated by a bound cyclin, CDKs perform a common biochemical reaction called phosphorylation that activates or inactivates target proteins to orchestrate coordinated entry into the next phase of the cell cycle. Different cyclin-CDK combinations determine the downstream proteins targeted. CDKs are constitutively expressed in cells whereas cyclins are synthesised at specific stages of the cell cycle, in response to various molecular signals.

General Mechanism of Cyclin-CDK Interaction

Upon receiving a pro-mitotic extracellular signal, G_1 cyclin-CDK complexes become active to prepare the cell for S phase, promoting the expression of transcription factors that in turn promote the expression of S cyclins and of enzymes required for DNA replication. The G_1 cyclin-CDK complexes also promote the degradation of molecules that function as S phase inhibitors by targeting them for ubiquitination. Once a protein has been ubiquitinated, it is targeted for proteolytic degradation by the proteasome. However, results from a recent study of E2F transcriptional dynamics at the single-cell level argue that the role of G1 cyclin-CDK activities, in particular cyclin D-CDK4/6, is to tune the timing rather than the commitment of cell cycle entry.

Active S cyclin-CDK complexes phosphorylate proteins that make up the pre-replication complexes assembled during G_1 phase on DNA replication origins. The phosphorylation serves two purposes: to activate each already-assembled pre-replication complex, and to prevent new complexes from forming. This ensures that every portion of the cell's genome will be replicated once and only once. The reason for prevention of gaps in replication is fairly clear, because daughter cells that are missing all or part of crucial genes will die. However, for reasons related to gene copy number effects, possession of extra copies of certain genes is also deleterious to the daughter cells.

Mitotic cyclin-CDK complexes, which are synthesized but inactivated during S and G_2 phases, promote the initiation of mitosis by stimulating downstream proteins involved in chromosome condensation and mitotic spindle assembly. A critical complex activated during this process is a ubiquitin ligase known as the anaphase-promoting complex (APC), which promotes degradation of structural proteins associated with the chromosomal kinetochore. APC also targets the mitotic cyclins for degradation, ensuring that telophase and cytokinesis can proceed.

Specific Action of Cyclin-CDK Complexes

Cyclin D is the first cyclin produced in the cell cycle, in response to extracellular signals (e.g. growth factors). Cyclin D binds to existing CDK4, forming the active cyclin D-CDK4 complex. Cyclin D-CDK4 complex in turn phosphorylates the retinoblastoma susceptibility protein (Rb). The hyperphosphorylated Rb dissociates from the E2F/DP1/Rb complex (which was bound to the E2F responsive genes, effectively "blocking" them from transcription), activating E2F. Activation of E2F results in transcription of various genes like cyclin E, cyclin A, DNA polymerase, thymidine kinase, etc. Cyclin E thus produced binds to CDK2, forming the cyclin E-CDK2 complex, which pushes the cell from G_1 to S phase (G_1/S, which initiates the G_2/M transition). Cyclin B-cdk1 complex activation causes breakdown of nuclear envelope and initiation of prophase, and subsequently, its deactivation causes the cell to exit mitosis.

A quantitative study of E2F transcriptional dynamics at the single-cell level by using engineered fluorescent reporter cells provided a quantitative framework for understanding the control logic of cell cycle entry, challenging the canonical textbook model. Genes that regulate the amplitude of E2F accumulation, such as Myc, determine the commitment into cell cycle and S phase entry. G1 cyclin-CDK activities are not the driver of cell cycle entry. Instead, they primarily tune the timing of E2F increase, thereby modulating the pace of cell cycle progression.

Inhibitors

Two families of genes, the *cip/kip* (*CDK interacting protein/Kinase inhibitory protein*) family and the INK4a/ARF (*In*hibitor of *K*inase 4/*A*lternative *R*eading *F*rame) family, prevent the progression of the cell cycle. Because these genes are instrumental in prevention of tumor formation, they are known as tumor suppressors.

The cip/kip family includes the genes p21, p27 and p57. They halt cell cycle in G_1 phase, by binding to, and inactivating, cyclin-CDK complexes. p21 is activated by p53 (which, in turn, is triggered by DNA damage e.g. due to radiation). p27 is activated by Transforming Growth Factor of β (TGF β), a growth inhibitor.

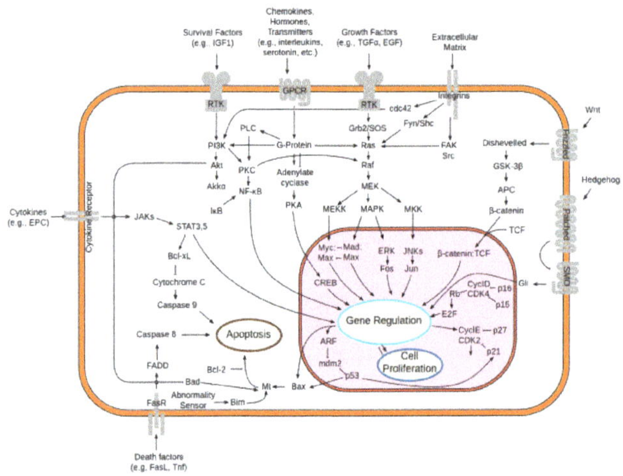

Overview of signal transduction pathways involved in apoptosis, also known as "programmed cell death".

The INK4a/ARF family includes p16^{INK4a}, which binds to CDK4 and arrests the cell cycle in G_1 phase, and p14^{ARF} which prevents p53 degradation.

Synthetic inhibitors of Cdc25 could also be useful for the arrest of cell cycle and therefore be useful as antineoplastic and anticancer agents.

Transcriptional Regulatory Network

Current evidence suggests that a semi-autonomous transcriptional network acts in concert with the CDK-cyclin machinery to regulate the cell cycle. Several gene expression studies in *Saccharomyces cerevisiae* have identified 800-1200 genes that change expression over the course of the cell cycle. They are transcribed at high levels at specific points in the cell cycle, and remain at lower levels throughout the rest of the cycle. While the set of identified genes differs between studies due to the computational methods and criteria used to identify them, each study indicates that a large portion of yeast genes are temporally regulated.

Many periodically expressed genes are driven by transcription factors that are also periodically expressed. One screen of single-gene knockouts identified 48 transcription factors (about 20% of all non-essential transcription factors) that show cell cycle progression defects. Genome-wide studies using high throughput technologies have identified the transcription factors that bind to the promoters of yeast genes, and correlating these findings with temporal expression patterns have allowed the identification of transcription factors that drive phase-specific gene expression. The expression profiles of these transcription factors are driven by the transcription factors that peak in the prior phase, and computational models have shown that a CDK-autonomous network of these transcription factors is sufficient to produce steady-state oscillations in gene expression).

Experimental evidence also suggests that gene expression can oscillate with the period seen in dividing wild-type cells independently of the CDK machinery. Orlando *et al.* used microarrays to measure the expression of a set of 1,271 genes that they identified as periodic in both wild type cells and cells lacking all S-phase and mitotic cyclins (*clb1,2,3,4,5,6*). Of the 1,271 genes assayed, 882 continued to be expressed in the cyclin-deficient cells at the same time as in the wild type cells, despite the fact that the cyclin-deficient cells arrest at the border between G_1 and S phase. However, 833 of the genes assayed changed behavior between the wild type and mutant cells, indicating that these genes are likely directly or indirectly regulated by the CDK-cyclin machinery. Some genes that continued to be expressed on time in the mutant cells were also expressed at different levels in the mutant and wild type cells. These findings suggest that while the transcriptional network may oscillate independently of the CDK-cyclin oscillator, they are coupled in a manner that requires both to ensure the proper timing of cell cycle events. Other work indicates that phosphorylation, a post-translational modification, of cell cycle transcription factors by Cdk1 may alter the localization or activity of the transcription factors in order to tightly control timing of target genes.

While oscillatory transcription plays a key role in the progression of the yeast cell cycle, the CDK-cyclin machinery operates independently in the early embryonic cell cycle. Before the midblastula transition, zygotic transcription does not occur and all needed proteins, such as the B-type cyclins, are translated from maternally loaded mRNA.

DNA Replication and DNA Replication Origin Activity

Analyses of synchronized cultures of *Saccharomyces cerevisiae* under conditions that prevent DNA replication initiation without delaying cell cycle progression showed that origin licensing decreases the expression of genes with origins near their 3' ends, revealing that downstream origins can regulate the expression of upstream genes. This confirms previous predictions from mathematical modeling of a global causal coordination between DNA replication origin activity and mRNA expression, and shows that mathematical modeling of DNA microarray data can be used to correctly predict previously unknown biological modes of regulation.

Checkpoints

Cell cycle checkpoints are used by the cell to monitor and regulate the progress of the cell cycle. Checkpoints prevent cell cycle progression at specific points, allowing verification of necessary phase processes and repair of DNA damage. The cell cannot proceed to the next phase until checkpoint requirements have been met. Checkpoints typically consist of a network of regulatory proteins that monitor and dictate the progression of the cell through the different stages of the cell cycle.

There are several checkpoints to ensure that damaged or incomplete DNA is not passed on to daughter cells. Three main checkpoints exist: the G_1/S checkpoint, the G_2/M checkpoint and the metaphase (mitotic) checkpoint.

G_1/S transition is a rate-limiting step in the cell cycle and is also known as restriction point. This is where the cell checks whether it has enough raw materials to fully replicate its DNA (nucleotide bases, DNA synthase, chromatin, etc.). An unhealthy or malnourished cell will get stuck at this checkpoint.

The G_2/M checkpoint is where the cell ensures that it has enough cytoplasm and phospholipids for two daughter cells. But sometimes more importantly, it checks to see if it is the right time to replicate. There are some situations where many cells need to all replicate simultaneously (for example, a growing embryo should have a symmetric cell distribution until it reaches the mid-blastula transition). This is done by controlling the G_2/M checkpoint.

The metaphase checkpoint is a fairly minor checkpoint, in that once a cell is in metaphase, it has committed to undergoing mitosis. However that's not to say it isn't important. In this checkpoint, the cell checks to ensure that the spindle has formed and that all of the chromosomes are aligned at the spindle equator before anaphase begins.

While these are the three "main" checkpoints, not all cells have to pass through each of these checkpoints in this order to replicate. Many types of cancer are caused by mutations that allow the cells to speed through the various checkpoints or even skip them altogether. Going from S to M to S phase almost consecutively. Because these cells have lost their checkpoints, any DNA mutations that may have occurred are disregarded and passed on to the daughter cells. This is one reason why cancer cells have a tendency to exponentially accrue mutations. Aside from cancer cells, many fully differentiated cell types no longer replicate so they leave the cell cycle and stay in G_0 until their death. Thus removing the need for cellular checkpoints. An alternative model of the cell cycle response to DNA damage has also been proposed, known as the postreplication checkpoint.

Checkpoint regulation plays an important role in an organism's development. In sexual reproduction, when egg fertilization occurs, when the sperm binds to the egg, it releases signalling factors that notify the egg that it has been fertilized. Among other things, this induces the now fertilized oocyte to return from its previously dormant, G_0, state back into the cell cycle and on to mitotic replication and division.

p53 plays an important role in triggering the control mechanisms at both G_1/S and G_2/M checkpoints. In addition to p53, checkpoint regulators are being heavily researched for their roles in cancer growth and proliferation.

Role in Tumor Formation

A disregulation of the cell cycle components may lead to tumor formation. As mentioned above, when some genes like the cell cycle inhibitors, RB, p53 etc. mutate, they may cause the cell to multiply uncontrollably, forming a tumor. Although the duration of cell cycle in tumor cells is equal to or longer than that of normal cell cycle, the proportion of cells that are in active cell division (versus quiescent cells in G_0 phase) in tumors is much higher than that in normal tissue. Thus there is a net increase in cell number as the number of cells that die by apoptosis or senescence remains the same.

The cells which are actively undergoing cell cycle are targeted in cancer therapy as the DNA is relatively exposed during cell division and hence susceptible to damage by drugs or radiation. This fact is made use of in cancer treatment; by a process known as debulking, a significant mass of the tumor is removed which pushes a significant number of the remaining tumor cells from G_0 to G_1 phase (due to increased availability of nutrients, oxygen, growth factors etc.). Radiation or chemotherapy following the debulking procedure kills these cells which have newly entered the cell cycle.

The fastest cycling mammalian cells in culture, crypt cells in the intestinal epithelium, have a cycle time as short as 9 to 10 hours. Stem cells in resting mouse skin may have a cycle time of more than 200 hours. Most of this difference is due to the varying length of G_1, the most variable phase of the cycle. M and S do not vary much.

In general, cells are most radiosensitive in late M and G_2 phases and most resistant in late S phase.

For cells with a longer cell cycle time and a significantly long G_1 phase, there is a second peak of resistance late in G_1.

The pattern of resistance and sensitivity correlates with the level of sulfhydryl compounds in the cell. Sulfhydryls are natural substances that protect cells from radiation damage and tend to be at their highest levels in S and at their lowest near mitosis.

Fluorescence Imaging of the Cell Cycle

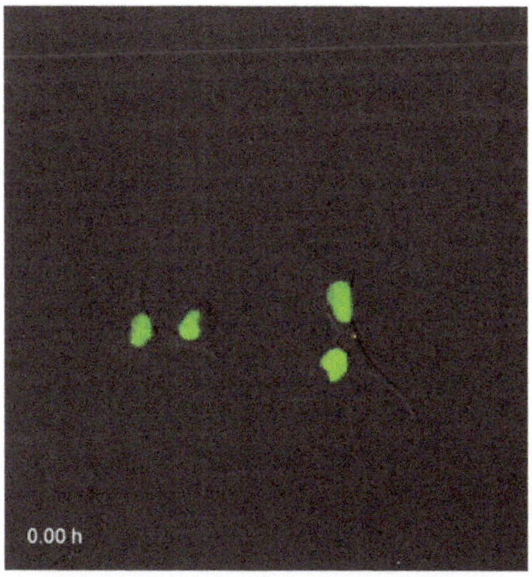

Fluorescent proteins visualize the cell cycle progression. IFP2.0-hGem(1/110) fluorescence is shown in green and highlights the S/G_2/M phases. smURFP-hCdtI(30/120) fluorescence is shown in red and highlights the G_0/G_1 phases.

Pioneering work by Atsushi Miyawaki and coworkers developed the fluorescent ubiquitination-based cell cycle indicator (FUCCI), which enables fluorescence imaging of the cell cycle. Originally, a green fluorescent protein, mAG, was fused to hGem(1/110) and an orange fluorescent protein (mKO_2) was fused to hCdt1(30/120). Note, these fusions are fragments that contain a nuclear localization signal and ubiquitination sites for degradation, but are not functional proteins. The green fluorescent protein is made during the S, G_2, or M phase and degraded during the G_0 or G_1 phase, while the orange fluorescent protein is made during the G_0 or G_1 phase and destroyed during the S, G_2, or M phase. A far-red and near-infrared FUCCI was developed using a cyanobacteria-derived fluorescent protein (smURFP) and a bacteriophytochrome-derived fluorescent protein (movie found at this link).

Cell Cycle Checkpoint

Cell cycle checkpoints are control mechanisms in eukaryotic cells which ensure proper division of the cell. Each checkpoint serves as a potential point along the cell cycle, during which the conditions of the cell are assessed, with progression through the various phases of the cell cycle occurring when favorable conditions are met. Currently, there are three known checkpoints: the G1 checkpoint, also known as the restriction or start checkpoint or (Major Checkpoint); the G2/M checkpoint; and the metaphase checkpoint, also known as the spindle checkpoint.

Background

All living organisms are products of repeated rounds of cell growth and division. During this process, known as the cell cycle, a cell duplicates its contents and then divides in two. The purpose of the cell cycle is to accurately duplicate each organisms' DNA and then divide the cell and its contents evenly between the two resulting cells. In eukaryotes, the cell cycle consists of four main stages: G1, during which a cell is metabolically active and continuously grows; S phase, during which DNA replication takes place; G2, during which cell growth continues and the cell synthesizes various proteins in preparation for division; and the M (mitosis) phase, during which the duplicated chromosomes (known as the sister chromatids) separate into two daughter nuclei, and the cell divides into two daughter cells, each with a full copy of DNA. Compared to the eukaryotic cell cycle, the prokaryotic cell cycle (known as binary fission) is relatively simple and quick: the chromosome replicates from the origin of replication, a new membrane is assembled, and the cell wall forms a septum which divides the cell into two.

As the eukaryotic cell cycle is a complex process, eukaryotes have evolved a network of regulatory proteins, known as the cell cycle control system, which monitors and dictates the progression of the cell through the cell cycle. This system acts like a timer, or a clock, which sets a fixed amount of time for the cell to spend in each phase of the cell cycle, while at the same time it also responds to information received from the processes it controls. The cell cycle checkpoints play an important role in the control system by sensing defects that occur during essential processes such as DNA replication or chromosome segregation, and inducing a cell cycle arrest in response until the defects are repaired. The main mechanism of action of the cell cycle checkpoints is through the regulation of the activities of a family of protein kinases known as the cyclin-dependent kinases (CDKs), which bind to different classes of regulator proteins known as cyclins, with specific cyclin-CDK complexes being formed and activated at different phases of the cell cycle. Those complexes, in turn, activate different downstream targets to promote or prevent cell cycle progression.

G_1 (Restriction) Checkpoint

The G1 checkpoint, also known as the restriction point in mammalian cells and the start point in yeast, is the point at which the cell becomes committed to entering the cell cycle. As the cell progresses through G1, depending on internal and external conditions, it can either delay G1, enter a quiescent state known as G0, or proceed past the restriction point. The decision to commit to a new round of cell division occurs when the cell activates cyclin-CDK-dependent transcription which promotes entry into S phase.

During early G1, the transcriptional repressors Rb (retinoblastoma), p107 and p130, known as pocket proteins, bind to the E2F transcription factors to prevent G1-to-S transition. Rb binds and represses activator E2F transcription factors (E2F1-3), while p107 and p130 bind E2F4 and E2F5 to form complexes which repress transcription of G1-to-S promoting factors. Upon the decision to progress past the G1 checkpoint, cyclin D levels rise, and cyclin D forms a complex with CDK4 and CDK6, which in turn phosphorylate the pocket proteins. Phosphorylation of the pocket proteins causes the release of their bound targets, thereby relieving the repression of the E2F1-3 activators and translocating repressors E2F4 and E2F5 from the nucleus to the cytoplasm. This results in the transcriptional activation of downstream targets, which promote the G1-to-S transition, including another cyclin, known as cyclin E, which forms a complex with CDK2. The formation of the cyclin E-CDK2 complex then promotes a positive feedback loop which creates an "all or nothing" switch from which the cell can not return. Following entry to S-phase and initiation of DNA replication, S-phase cyclin A, a transcriptional target of E2F1-3, forms a complex with CDK2 which phosphorylates E2F1-3 and prevents its ability to bind to DNA, thus forming a negative feedback loop. In another negative feedback loop, E2F1-3 promote the transcription of E2F6-8, which in turn repress G1-S transition.

When DNA damage occurs, or when the cell detects any defects which necessitate it to delay or halt the cell cycle in G1, arrest occurs through several mechanisms. The rapid response involves phosphorylation events that initiate with either kinase ATM (Ataxia telangiectasia mutated) or ATR (Ataxia Telangiectasia and Rad3 related), which act as sensors, depending on the type of damage. These kinases phosphorylate and activate the effector kinases Chk2 and Chk1, respectively, which in turn phosphorylate the phosphatase Cdc25A, thus marking it for ubiquitination and degradation. As Cdc25A activates the previously mentioned cyclin E-CDK2 complex by removing inhibitory phosphates from CDK2, in the absence of Cdc25A, cyclin E-CDK2 remains inactive, and the cell remains in G1. To maintain the arrest, another response is initiated, by which Chk2 or Chk1 phosphorylate p53, a tumor suppressor, and this stabilizes p53 by preventing it from binding Mdm2, a ubiquitin ligase which inhibits p53 by targeting it for degradation. The stable p53 then acts a transcriptional activator of several target genes, including p21, an inhibitor of the G1-to-S promoting complex cyclin E-CDK21. In addition, another mechanism by which p21 is activated is through the accumulation of p16 in response to DNA damage. p16 disrupts cyclin D-CDK4 complexes, thus causing the release of p21 from the complexes, which leads to the dephosphorylation and activation of Rb, which allows Rb to bind and inhibit E2F1-3, thus keeping the cell from transitioning to S phase. Recently, some aspects of this model have been disputed.

G_2 Checkpoint

Following the decision to enter the cell cycle and undergo division, the cell goes through S phase, in which it replicates its DNA, and, in most species, G2, in which it undergoes rapid growth and protein synthesis in preparation for mitosis, the process of cell division. The G2/M checkpoint, also known as the DNA damage checkpoint, ensures that the cell underwent all of the necessary changes during the S and G2 phases and is ready to divide.

The primary complex responsible for the transition from G2 to M is the Cyclin B-cdc2 (CDK1 homolog) complex. The activity of cdc2 is regulated directly by cyclins B and by the phosphatase cdc25. Prior to entry into mitosis, cdc2 is maintained in an inactive state by the kinases Wee1 and Myt1, which phosphorylate tyrosine residues on cdc2. As the cell progresses through G2 and reaches the G2/M transition, the kinase Plk1 phosphorylates Wee1, which targets Wee1 for degradation via the SCF ubiquitin ligase complex. An additional function of Plk1 is to activate Cdc25 through phosphorylation. The compound effect of Wee1 degradation and Cdc25 activation is the net removal of inhibitory phosphorylation from cdc2, which activates cdc2. Plk1 is activated at the G2/M transition by the Aurora A and Bora, which accumulate during G2 and form an activation complex. The Plk1-Cdc2-cdc25 complex then initiates a positive feedback loop which serves to further activate Cdc2, and in conjunction with an increase in cyclin B levels during G2, the resulting cdc2-cyclin B complexes then activate downstream targets which promote entry into mitosis.

The mechanisms by which mitotic entry is prevented in response to DNA damage are similar to those in the G1/S checkpoint. DNA damage triggers the activation of the aforementioned ATM/ATR pathway, in which ATM/ATR phosphorylate and activate the Chk1/Chk2 checkpoint kinases. Chk1/2 phosphorylate cdc25 which, in addition to being inhibited, is also sequestered in the cytoplasm by the 14-3-3 proteins. 14-3-3 are upregulated by p53, which, as previously mentioned, is activated by Chk1 and ATM/ATR. p53 also transactivates p21, and both p21 and the 14-3-3 in turn inhibit cyclin B-cdc2 complexes through the phosphorylation and cytoplasmic sequestering of cdc2. In addition, the inactivation of cdc25 results in its inability to dephosphorylate and activate cdc2. Finally, another mechanism of damage response is through the negative regulation of Plk1 by ATM/ATR, which in turn results in the stabilization of Wee1 and Myt1, which can then phosphorylate and inhibit cdc2, thus keeping the cell arrested in G2 until the damage is fixed.

Metaphase Checkpoint

The mitotic spindle checkpoint occurs at the point in metaphase where all the chromosomes should/have aligned at the mitotic plate and be under bipolar tension. The tension created by this bipolar attachment is what is sensed, which initiates the anaphase entry. To do this, the sensing mechanism ensures that the anaphase-promoting complex (APC/C) is no longer inhibited, which is now free to degrade cyclin B, which harbors a D-box (destruction box), and to break down securin. The latter is a protein whose function is to inhibit separase, which in turn cuts the cohesins, the protein composite responsible for cohesion of sister chromatids. Once this inhibitory protein is degraded via ubiquitination and subsequent proteolysis, separase then causes sister chromatid separation. After the cell has split into its two daughter cells, the cell enters G_1.

Cell Cycle Checkpoints and Cancer

DNA repair processes and cell cycle checkpoints have been intimately linked with cancer due to their functions regulating genome stability and cell progression, respectively. The precise molecular mechanisms that connect dysfunctions in these pathways to the onset of particular cancers are not well understood in most cases. The loss of ATM has been shown to precede lymphoma development presumably due to excessive homologous recombination, leading to high genomic instability. Disruption of Chk1 in mice led significant misregulation of cell cycle checkpoints, an acculumation of DNA damage, and an increased incidence of tumorigenesis. Perhaps most famously, single mutant inheritance of BRCA1 or BRCA2 predisposes women toward breast and ovarian cancers. BRCA1 is known to be required for S and G2/M transitions, and is involved in the cellular response to DNA damage. BRCA2 is believed to be involved in homologous recombination and regulating the S-phase checkpoint, and mutations of deficiencies in BRCA2 are strongly linked to tumorigenesis.

Nondisjunction

Nondisjunction is the failure of homologous chromosomes or sister chromatids to separate properly during cell division. There are three forms of nondisjunction: failure of a pair of homologous chromosomes to separate in meiosis I, failure of sister chromatids to separate during meiosis II, and failure of sister chromatids to separate during mitosis. Nondisjunction results in daughter cells with abnormal chromosome numbers (aneuploidy).

Calvin Bridges and Thomas Hunt Morgan are credited with discovering nondisjunction in *Drosophila melanogaster* sex chromosomes in the spring of 1910, while working in the Zoological Laboratory of Columbia University.

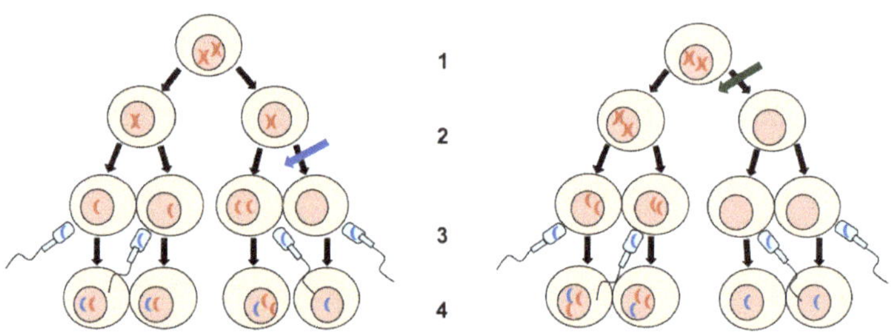

1. Meiosis I 2. Meiosis II 3. Fertilization 4. Zygote The left image at the blue arrow is nondisjunction taking place during meiosis II. The right image at the green arrow is nondisjunction taking place during meiosis I. Nondisjunction is when chromosomes fail to separate normally resulting in a gain or loss of chromosomes.

Types

In general, nondisjunction can occur in any form of cell division that involves ordered distribution of chromosomal material. Higher animals have three distinct forms of such cell divisions: Meiosis I and meiosis II are specialized forms of cell division occurring during generation of gametes (eggs and sperm) for sexual reproduction, mitosis is the form of cell division used by all other cells of the body.

Meiosis II

Ovulated eggs become arrested in metaphase II until fertilization triggers the second meiotic division. Similar to the segregation events of mitosis, the pairs of sister chromatids resulting from the separation of bivalents in meiosis I are further separated in anaphase of meiosis II. In oocytes, one sister chromatid is segregated into the second polar body, while the other stays inside the egg. During spermatogenesis, each meiotic division is symmetric such that each primary spermatocyte gives rise to 2 secondary spermatocytes after meiosis I, and eventually 4 spermatids after meiosis II. Meiosis II-nondisjunction may also result in aneuploidy syndromes, but only to a much smaller extent than do segregation failures in meiosis I.

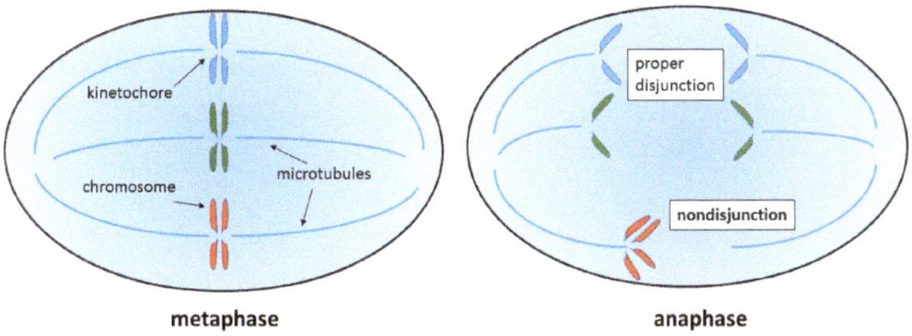

Nondisjunction of sister chromatids during mitosis:

Left: Metaphase of mitosis. Chromosome line up in the middle plane, the mitotic spindle forms and the kinetochores of sister chromatids attach to the microtubules.

Right: Anaphase of mitosis, where sister chromatids separate and the microtubules pull them in opposite directions.

The chromosome shown in *red* fails to separate properly, its sister chromatids stick together and get pulled to the same side, resulting in mitotic nondisjunction of this chromosome.

Mitosis

Division of somatic cells through mitosis is preceded by replication of the genetic material in S phase. As a result, each chromosome consists of two sister chromatids held together at the centromere. In the anaphase of mitosis, sister chromatids separate and migrate to opposite cell poles before the cell divides. Nondisjunction during mitosis leads to one daughter receiving both sister chromatids of the affected chromosome while the other gets none. This is known as a chromatin bridge or an anaphase bridge. Mitotic nondisjunction results in somatic mosaicism, since only daughter cells originating from the cell where the nondisjunction event has occurred will have an abnormal number of chromosomes. Nondisjunction during mitosis can contribute to the development of some forms of cancer, e.g. retinoblastoma. Chromosome nondisjunction in mitosis can be atrributed to the inactivation of topoisomerase II, condensin, or separase. Meiotic nondisjunction has been well studied in *Saccharomyces cerevisiae*. This yeast undergoes mitosis similarly to other eukaryotes. Chromosome bridges occur when sister chromatids are held together post replication by DNA-DNA topological entanglement and the cohesion complex. During anaphase, cohesion is cleaved by separase. Topoisomerase II and condensin are responsible for removing catenations.

Molecular Mechanisms

Central Role of the Spindle Assembly Checkpoint

The spindle assembly checkpoint (SAC) is a molecular safe-guarding mechanism that governs proper chromosome segregation in eukaryotic cells. SAC inhibits progression into anaphase until all homologous chromosomes (bivalents) are properly aligned to the spindle apparatus. Only then, SAC releases its inhibition of the anaphase promoting complex (APC), which in turn irreversibly triggers progression through anaphase.

Sex-specific Differences in Meiosis

Surveys of cases of human aneuploidy syndromes have shown that most of them are maternally derived. This raises the question: Why is female meiosis more error prone? The most obvious difference between female oogenesis and male spermatogenesis is the prolonged arrest of oocytes in late stages of prophase I for many years up to several decades. Male gametes on the other hand quickly go through all stages of meiosis I and II. Another important difference between male and female meiosis concerns the frequency of recombination between homologous chromosomes: In the male, almost all chromosome pairs are joined by at least one crossover, while more than 10% of human oocytes contain at least one bivalent without any crossover event. Failures of recombination or inappropriately located crossovers have been well documented as contributors to the occurrence of nondisjunction in humans.

Age-related Loss of Cohesin Ties

Due to the prolonged arrest of human oocytes, weakening of cohesive ties holding together chromosomes and reduced activity of the SAC may contribute to maternal age-related errors in segregation control. The cohesin complex is responsible for keeping together sister chromatids and provides binding sites for spindle attachment. Cohesin is loaded onto newly replicated chromosomes in oogonia during fetal development. Mature oocytes have only limited capacity for reloading cohesin after completion of S phase. The prolonged arrest of human oocytes prior to completion of meiosis I may therefore result in considerable loss of cohesin over time. Loss of cohesin is assumed to contribute to incorrect microtubule-kinetochore attachment and chromosome segregation errors during meiotic divisions.

Consequences

The result of this error is a cell with an imbalance of chromosomes. Such a cell is said to be aneuploid. Loss of a single chromosome (2n-1), in which the daughter cell(s) with the defect will have one chromosome missing from one of its pairs, is referred to as a monosomy. Gaining a single chromosome, in which the daughter cell(s) with the defect will have one chromosome in addition to its pairs is referred to as a trisomy. In the event that an aneuploidic gamete is fertilized, a number of syndromes might result.

Monosomy

The only known survivable monosomy in humans is Turner syndrome, where the affected individual is monosomic for the X chromosome. Other monosomies are usually lethal during early fetal development, and survival is only possible if not all the cells of the body are affected in case of a mosaicism, or if the normal number of chromosomes is restored via duplication of the single monosomic chromosome ("chromosome rescue").

Turner Syndrome (X Monosomy) (45, Xo)

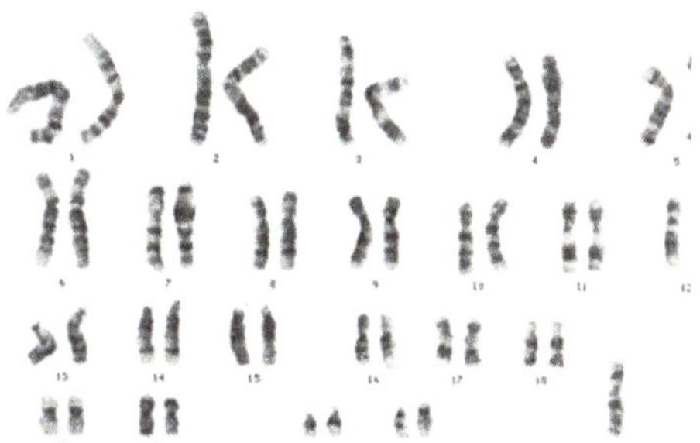

Karyotype of X monosomy (Turner syndrome) This condition is characterized by the presence of only one X chromosome and no Y chromosome (see bottom right corner).

Complete loss of an entire X chromosome accounts for about half the cases of Turner syndrome. The importance of both X chromosomes during embryonic development is underscored by the observation that the overwhelming majority (>99%) of fetuses with only one X chromosome (karyotype 45, Xo) are spontaneously aborted.

Autosomal Trisomy

The term autosomal trisomy means that a chromosome other than the sex chromosomes X and Y is present in 3 copies instead of the normal number of 2 in diploid cells.

Down Syndrome (Trisomy 21)

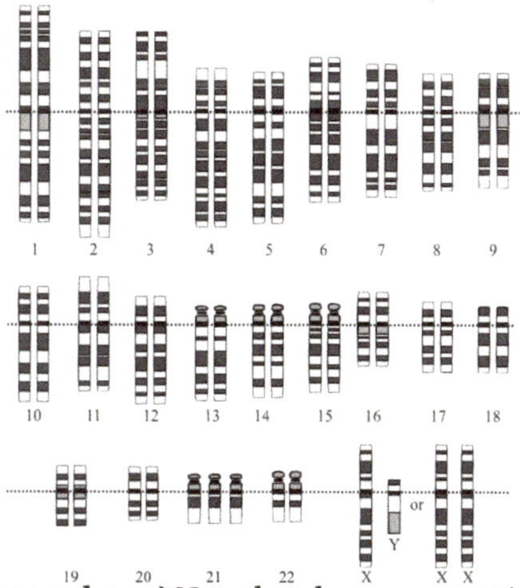

Karyotype of trisomy 21 (Down syndrome) Note that chromosome 21 is present in 3 copies, while all other chromosomes show the normal diploid state with 2 copies. Most cases of trisomy of chromosome 21 are caused by a nondisjunction event during meiosis I.

Down syndrome, a trisomy of chromosome 21, is the most common anomaly of chromosome number in humans. The majority of cases results from nondisjunction during maternal meiosis I. Trisomy occurs in at least 0.3% of newborns and in nearly 25% of spontaneous abortions. It is the leading cause of pregnancy wastage and is the most common known cause of mental retardation. It is well documented that advanced maternal age is associated with greater risk of meiotic nondisjunction leading to Down syndrome. This may be associated with the prolonged meiotic arrest of human oocytes potentially lasting for more than four decades.

Edwards Syndrome (Trisomy 18) and Patau Syndrome (Trisomy 13)

Human trisomies compatible with life birth, other than Down syndrome (trisomy 21), are Edwards syndrome (trisomy 18), and Patau syndrome (trisomy 13). Complete trisomies of other chromosomes are usually not viable and represent a relatively frequent cause of miscarriage. Only in rare cases of a mosaicism, the presence of a normal cell line, in addition to the trisomic cell line, may support the development of a viable trisomy of the other chromosomes.

Sex Chromosome Aneuploidy

The term *sex chromosome aneuploidy* summarizes conditions with an abnormal number of sex chromosomes, i.e. other than XX (female) or XY (male). Formally, X chromosome monosomy can also be classified as a form of sex chromosome aneuploidy.

Klinefelter Syndrome (47, Xxy)

Klinefelter syndrome is the most common sex chromosome aneuploidy in humans. It represents the most frequent cause of hypogonadism and infertility in men. Most cases are caused by nondisjunction errors in paternal meiosis I. About eighty percent of individuals with this syndrome have one extra X chromosome resulting in the karyotype XXY. The remaining cases have either multiple additional sex chromosomes (48,XXXY; 48,XXYY; 49,XXXXY), mosaicism (46,XY/47,XXY), or structural chromosome abnormalities.

XYY Male (47, XYY)

The incidence of XYY syndrome is approximately 1 in 800-1000 male births. Many cases remain undiagnosed because of their normal appearance and fertility, and the absence of severe symptoms. The extra Y chromosome is usually a result of nondisjunction during paternal meiosis II.

Trisomy X (47,XXX)

Trisomy X is a form of sex chromosome aneuploidy where females have three instead of two X chromosomes. Most patients are only mildly affected by neuropsychological and physical symptoms. Studies examining the origin of the extra X chromosome observed that about 58-63% of cases were caused by nondisjunction in maternal meiosis I, 16-18% by nondisjunction in maternal meiosis II, and the remaining cases by post-zygotic, i.e. mitotic, nondisjunction.

Uniparental Disomy

Uniparental disomy denotes the situation where both chromosomes of a chromosome pair are inherited from the same parent and are therefore identical. This phenomenon most likely is the result of a pregnancy that started as a trisomy due to nondisjunction. Since most trisomies are lethal, the fetus only survives because it loses one of the three chromosomes and becomes disomic. Uniparental disomy of chromosome 15 is, for example, seen in some cases of Prader-Willi syndrome and Angelman syndrome.

Mosaicism Syndromes

Mosaicism syndromes can be caused by mitotic nondisjunction in early fetal development. As a consequence, the organism evolves as a mixture of cell lines with differing ploidy (number of chromosomes). Mosaicism may be present in some tissues, but not in others. Affected individuals may have a patchy or assymmetric appearance. Examples of mosaicism syndromes include Pallister-Killian syndrome and Hypomelanosis of Ito.

Mosaicism in Malignant Transformation

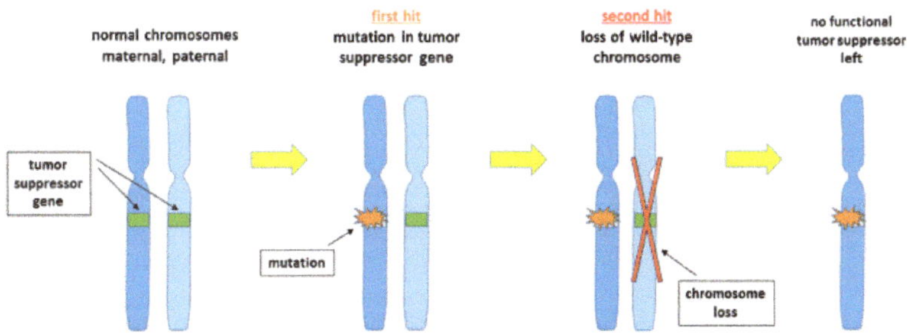

Loss of a tumor suppressor gene locus according to the two-hit model:

In the first hit, the tumor suppressor gene on one of the two chromosomes is affected by a mutation that makes the gene product non-functional. This mutation may arise spontaneously as a DNA replication error or may be induced by a DNA damaging agent. The second hit removes the remaining wild-type chromosome, for example through a mitotic nondisjunction event. There are several other potential mechanisms for each of the two steps, for example an additional mutation, an unbalanced translocation, or a gene deletion by recombination. As a result of the double lesion, the cell may become malignant because it is no longer able to express the tumor suppressor protein.

Development of cancer often involves multiple alterations of the cellular genome (Knudson hypothesis). Human retinoblastoma is a well studied example of a cancer type where mitotic nondisjunction can contribute to malignant transformation: Mutations of the RB1 gene, which is located on chromosome 13 and encodes the tumor suppressor retinoblastoma protein, can be detected by cytogenetic analysis in many cases of retinoblastoma. Mutations of the RB1 locus in one copy of chromosome 13 are sometimes accompanied by loss of the other wild-type chromosome 13 through mitotic nondisjunction. By this combination of lesions, affected cells completely lose expression of functioning tumor suppressor protein.

Diagnosis

Preimplantation Genetic Diagnosis

Pre-implantation genetic diagnosis (PGD or PIGD) is a technique used to identify genetically normal embryos and is useful for couples who have a family history of genetic disorders. This is an option for people choosing to procreate through IVF. PGD is considered difficult due to it being both time consuming and having success rates only comparable to routine IVF.

Karyotyping

Karyotyping involves performing an amniocentesis in order to study the cells of an unborn fetus. Light microscopy can be used to visually determine if aneuploidy is an issue.

Polar Body Diagnosis

Polar body diagnosis (PBD) can be use to detect maternally derived chromosomal aneuploidies as well as translocations in oocytes. The advantage of PBD over PGD is that it can be accomplished in a short amount of time. This is accomplished through zona drilling or laser drilling.

Blastomere Biopsy

Blastomere biopsy is a technique in which blastomeres are removed from the zona pellucida. It is commonly used to detect aneuploidy. Genetic analysis is conducted once the procedure is complete. Additional studies are needed to assess the risk associated with the procedure.

Lifestyle/Environmental Hazards

Exposure of spermatozoa to lifestyle, environmental and/or occupational hazards may increase the risk of aneuploidy. Cigarette smoke is a known aneugen (aneuploidy inducing agent). It is associated with increases in aneuploidy ranging from 1.5 to 3.0-fold. Other studies indicate factors such as alcohol consumption, occupational exposure to benzene, and exposure to the insecticides fenvalerate and carbaryl also increase aneuploidy.

Mitogen

A mitogen is a chemical substance that encourages a cell to commence cell division, triggering mitosis. A mitogen is usually some form of a protein. Mitogenesis is the induction (triggering) of mitosis, typically via a mitogen. Mitogens trigger signal transduction pathways in which mitogen-activated protein kinase (MAPK) is involved, leading to mitosis.

Use in Immunology

Lymphocytes can enter mitosis when they are activated by mitogens or antigens. B cells specifically can divide when they encounter an antigen matching their immunoglobulin. T cells undergo mitosis when stimulated by mitogens to produce small lymphocytes that are then responsible for the production of lymphokines, which are substances that modify the host organism to improve its immunity. B cells, on the other hand, divide to produce plasma cells when stimulated by mitogens, which then produce immunoglobulins, or antibodies. Mitogens are often used to stimulate lymphocytes and thereby assess immune function. The most commonly used mitogens in clinical laboratory medicine are:

Name	Acts upon T cells?	Acts upon B cells?
phytohaemagglutinin (PHA)	yes	no
concanavalin A (conA)	yes	no
lipopolysaccharide (LPS)	no	yes
pokeweed mitogen (PWM)	yes	yes

Lipopolysaccharide toxin from gram-negative bacteria is thymus-independent. They may directly activate B cells, regardless of their antigenic specificity. Plasma cells are terminally differentiated and, therefore, cannot undergo mitosis. Memory B cells can proliferate to produce more memory cells or plasma B cells. This is how the mitogen works, that is, by inducing mitosis in memory B cells to cause them to divide, with some becoming plasma cells.

Mitogens in Human Physiology

Insulin-like Growth Factor 1 mediates the major growth-promoting effect of Human

Growth Hormone as a paracrine agent at growth plates in the skeletal system. Another example for mitogen agent is G-CSF.

Other Uses

Mitogen-activated protein kinase (MAPK) pathways can induce enzymes such as the COX-2 enzyme. MAPK pathways may also play a role in the regulation of PTGS2.

Mitotic Cell Rounding

Mitotic cell rounding is a shape change that occurs in most animal cells that undergo mitosis. Cells abandon the spread or elongated shape characteristic of interphase and contract into a spherical morphology during mitosis. The phenomena is seen both in artificial cultures in vitro and naturally forming tissue in vivo.

Early Observations

In 1935, one of the first published accounts of mitotic rounding in live tissue described cell rounding in the pseudostratified epithelium of the mammalian neural tube. Sauer noticed that cells in mitosis rounded up to the apical, or luminal, surface of the columnar epithelium before dividing and returning to their elongated morphology.

Significance

For a long time it was not clear why cells became round in mitosis. Recent studies in the epithelia and epidermis of various organisms, however, show that mitotic cell rounding might serve several important functions.

- Firstly, mitotic cell rounding in combination with maintenance of apical cell-cell junctions appears to be necessary for correct mitotic spindle alignment, so that daughter cells divide parallel to the tissue plane, thus sharing apical surface to maintain tissue homeostasis. Failure to achieve this may result in mislocalization of one daughter cell to the basal region on the tissue layer and clearance via apoptotic cell death.

- Secondly, mitotic rounding has been proposed to be a driver for morphological events during tissue development. Examples include epithelial invagination of the Drosophila melanogaster tracheal placode and the anisotropic shape and growth of the inner ear lumen in Zebrafish.

- Thirdly, mitotic rounding has been shown to be important to generate sufficient space and appropriate geometry for proper mitotic spindle function, which is necessary for timely and accurate progression through mitosis.

Thus mitotic cell rounding is involved in tissue organization and homeostasis.

Mechanisms

To understand the physical mechanisms of how cells round up in mitosis researchers have conducted mechanical measurements with cultured cells in vitro. The forces that drive cell rounding have recently been characterized by researchers from the groups of Professors Tony Hyman and Daniel Muller, who used flat atomic force microscopycantilevers to constrain mitotic cells and measure the response force. More than 90% of the forces are generated by the collective activity of myosin IImolecular motors in the actin cortex. As a result, the surface tension and effective stiffness of the actin cortex increase as has been consistently observed in mitotic cells. This in turn yields an increase in intracellularhydrostatic pressure due to the Law of Laplace, which relates surface tension of a fluid interface to the differential pressure sustained across that interface. The increase in hydrostatic pressure is important because it produces the outward force necessary to push and rounds up against external objects or impediments, such as flexible cantilever or soft gel (in vitro examples), or surrounding extracellular matrix and neighboring cells (in vivo examples). In HeLa cellsin vitro, the force generated by a half-deformed mitotic cell is on the order of 50 to 100 nanonewtons. Internal hydrostatic pressure has been measured to increase from below 100 pascals in interphase to 3 to 10 fold that in mitosis.

In similar in vitro experiments, it was found that the threshold forces required to prevent mitosis are in excess of 100 nN. At threshold forces the cell suffers a loss of cortical F-actin uniformity, which further amplifies the susceptibility to applied force. These effects potentiate distortion of cell dimensions and subsequent perturbation of mitotic progression via spindle defects.

Release of stable focal adhesions is another important aspect of mitotic rounding. Cells that are genetically perturbed to manifest constitutively active adhesion regulators are unable to properly remodel their focal adhesions and facilitate the generation of a uniform actomyosin cortex. Overall, the biochemical events governing the morphological and mechanical changes in mitotic cells are orchestrated by the mitotic master regulator Cdk1.

References

- Griffiths, Anthony J.F.; Wessler, Susan R.; Carroll, Sean B.; Doebley, John (2012). Introduction to Genetic Analysis (10 ed.). New York: W.H. Freeman and Company. p. 35. ISBN 978-1-4292-2943-2.
- Griffiths JF, Gelbart WM, Lewontin RC, Wessler SR, Suzuki DT, Miller JH (2005). Introduction to Genetic Analysis. New York: W.H. Freeman and Co. pp. 34–40, 473–476, 626–629. ISBN 0-7167-4939-4.
- Cooper GM (2000). "Chapter 14: The Eukaryotic Cell Cycle". The cell: a molecular approach (2nd ed.). Washington, D.C: ASM Press. ISBN 0-87893-106-6.

- Robbins, Stanley L; Cotran, Ramzi S (2004). Vinay Kumar; Abul K Abbas; Nelson Fausto, eds. Pathological Basis of Disease. Elsevier. ISBN 81-8147-528-3.
- Norbury C (1995). "Cdk2 protein kinase (vertebrates)". In Hardie, D. Grahame; Hanks, Steven. Protein kinase factsBook. Boston: Academic Press. p. 184. ISBN 0-12-324719-5.
- Karp, Gerald (2005). Cell and Molecular Biology: Concepts and Experiments (4th ed.). Hoboken, New Jersey: John Wiley and Sons. pp. 598–9. ISBN 0-471-16231-0.
- Bacino, C.A.; Lee, B. (2011). "Chapter 76: Cytogenetics". In Kliegman, R.M.; Stanton, B.F.; St. Geme, J.W.; Schor, N.F.; Behrman, R.E. Nelson Textbook of Pediatrics, 19th Edition (19th ed.). Philadelphia: Saunders. pp. 394–413. ISBN 9781437707557.
- Gleason, [edited by] H. William Taeusch, Roberta A. Ballard, Christine A. (2005). Avery's diseases of the newborn (8th ed.). Philadelphia, Pa.: W.B. Saunders. ISBN 0721693474.
- LeMaire-Adkins, Renee. "Lack of Checkpoint Control at the Metaphase/Anaphase Transition: A Mechanism of Meiotic Nondisjunction in Mammalian Females". NCBI US National Library of Medicine. National Institute of Health. Retrieved 21 April 2016.
- Quammen, David (April 2008). "Contagious cancer: The evolution of a killer". Harper's. 316 (1895): 42. Retrieved 24 September 2012.

6

Phases of Cell Cycle and Mitosis

The period between mitotic divisions, known as G1 phase, S phase and G2 phase, is an interphase. Mitosis is a continuous process, which is divided into five stages which are prophase, prometaphase, metaphase, anaphase and telophase. The major categories of mitosis and cell cycle are dealt with great details in the chapter.

Interphase

Interphase is the phase of the cell cycle in which a typical cell spends most of its life. During this phase, the cell copies its DNA in preparation for mitosis. Interphase is the 'daily living' or metabolic phase of the cell, in which the cell obtains nutrients and metabolizes them, grows, reads its DNA, and conducts other "normal" cell functions. The majority of eukaryotic cells spend most of their time in interphase. This phase was formerly called the resting phase. However, interphase does not describe a cell that is merely resting; rather, the cell is actively living, and preparing for later cell division, so the name was changed. A common misconception is that interphase is the first stage of mitosis. However, since mitosis is the division of the nucleus, prophase is actually the first stage.

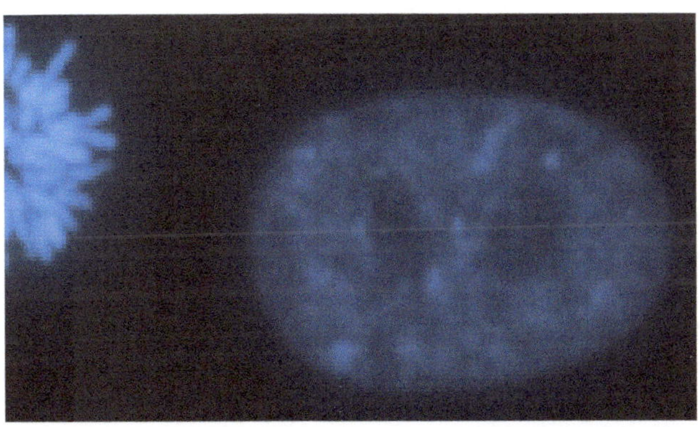

An image of the nucleus of a cell (HT1080) currently in interphase (likely G1). Note: Cytoplasm of this cell or the neighboring cell is not visible (top-left), which is currently in the telophase of mitosis. Image taken using an optical microscope and DAPI staining of DNA.

In interphase, the cell gets itself ready for mitosis or meiosis. Somatic cells, or normal diploid cells of the body, go through mitosis in order to reproduce themselves through cell division, whereas diploid germ cells (i.e., primary spermatocytes and primary oocytes) go through meiosis in order to create haploid gametes (i.e., sperm and ova) for the purpose of sexual reproduction. Chromosomes are copied.

Stages of Interphase

There are three stages of interphase, with each phase ending when a cellular checkpoint checks the accuracy of the stage's completion before proceeding to the next. The stages of interphase are:

- G_1 (Gap 1), in which the cell grows and functions normally. During this time, a high amount of protein synthesis occurs and the cell grows (to about double its original size) - more organelles are produced and the volume of the cytoplasm increases. If the cell is not to divide again, it will enter G_0.
- Synthesis (S), in which the cell duplicates its DNA (via semiconservative replication).
- G_2 (Gap 2), in which the cell resumes its growth in preparation for division. The mitochondria divide and the cell continues to grow until mitosis begins. In plants, chloroplasts also divide during G2.
- In addition, some cells that do not divide often or ever, enter a stage called G_0 (Gap zero), which is either a stage separate from interphase or an extended G_1.

The duration of time spent in interphase and in each stage of interphase is variable and depends on both the type of cell and the species of organism it belongs to. Most cells of adult mammals spend about 20 hours in interphase; this accounts for about 90% of the total time involved in cell division. Interphase includes G1, S, and G2 phases. Mitosis and cytokinesis, however, are separate from interphase.

Interphase within Sequences of Cellular Processes

Interphase and the Cell Cycle

When G_2 is completed, the cell enters a relatively brief period of nuclear and cellular division, composed of mitosis and cytokinesis, respectively. After the successful completion of mitosis and cytokinesis, both resulting daughter cells re-enter G_1 of interphase.

In the cell cycle, interphase is preceded by telophase and cytokinesis of the M phase. In alternative fashion, interphase is sometimes interrupted by G_0 phase, which, in some circumstances, may then end and be followed by the remaining stages of interphase. After the successful completion of the G_2 checkpoint, the final checkpoint in interphase, the cell proceeds to prophase, or in plants to preprophase, which is the first stage of mitosis.

G_0 phase is viewed as either an extended G_1 phase where the cell is neither dividing nor preparing to divide and or as a distinct quiescent stage which occurs outside of the cell cycle.

Interphase and Other Cellular Processes

In gamete production interphase is succeeded by meiosis. In programmed cell death, interphase is followed or preempted by apoptosis.

Phases of Interphase

G_1 phase

The G_1 phase, or Gap 1 phase, is the first of four phases of the cell cycle that takes place in eukaryotic cell division. In this part of interphase, the cell synthesizes mRNA and proteins in preparation for subsequent steps leading to mitosis. G_1 phase ends when the cell moves into the S phase of interphase.

Overview

G_1 phase together with the S phase and G_2 phase comprise the long growth period of the cell cycle called interphase that takes place before cell division in mitosis (M phase).

During G_1 phase, the cell grows in size and synthesizes mRNA and proteins (Known as histones) that are required for DNA synthesis. Once the required proteins and growth are complete, the cell enters the next phase of the cell cycle, S phase. The duration of each phase, including the G_1 phase, is different in many different types of cells. In human somatic cells, the cell cycle lasts about 18 hours, and the G_1 phase takes up about 1/3 of that time. However, in Xenopus embryos, sea urchin embryos, and Drosophila embryos, the G_1 phase is barely existent and is defined as the gap, if one exists, between the end of mitosis and the S phase.

G_1 phase and the other subphases of the cell cycle may be affected by limiting growth factors such as nutrient supply, temperature, and room for growth. Sufficient nucleotides and amino acids must be present in order to synthesize mRNA and proteins. Physiological temperatures are optimal for cell growth. In humans, the normal physiological temperature is around 37 °C (98.6 °F).

G_1 phase is particularly important in the cell cycle because it determines whether a cell commits to division or to leaving the cell cycle. If a cell is signaled to remain undivided, instead of moving onto the S phase, it will leave the G_1 phase and move into a state of dormancy called the G_0 phase. Most nonproliferating vertebrate cells will enter the G_0 phase.

Regulation of G_1 Phase

Within the cell cycle, there is a stringent set of regulations known as the cell cycle control system that controls the timing and coordination of the phases to ensure a correct order of events. Biochemical triggers known as cyclin-dependent kinases

(Cdks) switch on cell cycles events at the corrected time and in the correct order to prevent any mistakes.

There are three checkpoints in the cell cycle: the G_1/S Checkpoint or the Start checkpoint in yeast; the G_2/M checkpoint; and the spindle checkpoint.

Biochemical Regulators in G_1 Phase

During G_1 phase, the G_1/S cyclin activity rises significantly near the end of the G_1 phase.

Complexes of cyclin that are active during other phases of the cell cycle are kept inactivated to prevent any cell-cycle events from occurring out of order. Three methods of preventing Cdk activity are found in G_1 phase: inhibitory gene regulatory proteins suppress major cyclin genes from expression (*the way this is phrased it is unclear whether it is transcription of the gene or translation of the mRNA gene product that is being regulated); anaphase-promoting complex (APC) is activated, which targets and degrades S and M cyclins (but not G_1/S cyclins); and a high concentration of Cdk inhibitors is found during G_1 phase.

The Restriction Point

The Restriction Point (R) in the G_1 phase is different from a checkpoint because it does not determine whether cell conditions are ideal to move on to the next phase, but it changes the course of the cell. After a vertebrate cell has been in the G_1 phase for about three hours, the cell enters a restriction point in which it is decided whether the cell will move forward with the G_1 phase or move into the dormant G_0 phase.

This point also separates two halves of the G_1 phase; the post-mitotic and pre-mitotic phases. Between the beginning of the G_1 phase (which is also after mitosis has occurred) and R, the cell is known as being in the G_1-pm subphase, or the post-mitotic phase. After R and before S, the cell is known as being in G_1-ps, or the pre S phase interval of the G_1 phase.

In order for the cell to continue through the G_1-pm, there must be a high amount of growth factors and a steady rate of protein synthesis, otherwise the cell will move into G_0 phase.

Conflicting Research

Some authors will say that the Restriction Point and the G_1/S Checkpoint are one and the same, but more recent studies have argued that there are two different points in the G_1 phase that check the progression of the cell. The first restriction point is growth-factor dependent and determines whether the cell moves into the G_0 phase, while the second checkpoint is nutritionally-dependent and determines whether the cell moves into the S phase. Some of the confusion between researchers has been attributed to the study of mammalian vs. yeast cells.

The G_1/S Checkpoint

The G_1/S checkpoint is the point between G_1 phase and the S phase in which the cell is cleared for progression into the S phase. Reasons the cell would not move into the S phase include insufficient cell growth, damaged DNA, or other preparations have not been completed.

At the G_1/S checkpoint, formation of the G_1/S cyclin with Cdk to form a complex commits the cell to a new division cycle. These complexes then activate S-Cdk complexes that move forward with DNA replication in the S phase. Concurrently, anaphase-promoting complex (APC) activity decreases significantly, allowing S and M cyclins to become activated.

If a cell does not clear to pass through to the S phase, it enters the dormant G_0 phase in which there is no cellular growth or division.

The G_1 Phase in Cancer

Many sources have linked irregularities in the G_1 phase or the G_1/S checkpoint to uncontrolled growth of tumors. In these cases where the G_1 phase is affected, it is generally because gene regulatory proteins of the E2F family have become unrestrained and increase G_1/S cyclin gene expression, leading to uncontrolled cell-cycle entry.

However, the cure for some forms of cancer also lies in the G_1 phase of the cell cycle. Many cancers including breast and skin cancers have been prevented from proliferating by causing the tumor cells to enter G_1 cell cycle arrest, preventing the cells from dividing and spreading.

S Phase

S-phase (synthesis phase) is the part of the cell cycle in which DNA is replicated, occurring between G_1 phase and G_2 phase. Precise and accurate DNA replication is necessary to prevent genetic abnormalities which often lead to cell death or disease. Due to the importance, the regulatory pathways that govern this event in eukaryotes are highly conserved. This conservation makes the study of S-phase in model organisms such as *Xenopus laevis* embryos and budding yeast relevant to higher organisms.

S Phase Regulation

The G1/S transition is a major checkpoint in the regulation of the cell cycle. S Phase: To produce two similar daughter cells, the complete DNA instructions in the cell must be duplicated. DNA replication occurs during this S (synthesis) phase. Depending on levels of nutrients, energy and external factors, cells must decide to enter the cell cycle or move into a non-dividing state known as G_0 phase. This transition, as with all of the major checkpoint transitions in the cell cycle, is signaled by cyclins and cyclin dependent kinase (CDKs). The pulse of G1/S cyclins causes CLN3-Cdk1 to activate Cln1/2, (Start point (yeast)) as well as Clb5/6 at the initiation of S-phase. This pathway contains 2 positive feedback loops, allowing for rapid, unidirectional movement into S-phase. Redundant pathways like this are not uncommon because they allow for tuning the output of the system and often lead to faster genetic evolution.

DNA Replication

The major event in S-phase is DNA replication. The goal of this process is to create exactly two identical semi-conserved chromosomes. The cell prevents more than one replication from occurring by loading pre-replication complexes onto the DNA at replication origins during G_1 phase which are dismantled in S-phase as replication begins. In synthesis, the enzyme helicase unwinds the DNA double helix, and the enzyme DNA polymerase re-binds free-floating nucleotides to the separate DNA single strands in accordance with the complementary base pairing rule. DNA synthesis can occur as fast as 2000 nucleotides/second and must create fewer than 2 errors (I.e., wrong bases) in 10^{10} nucleotide additions.

DNA Damage

Damage to DNA is detected and fixed during S-phase. When the replication fork comes upon damaged DNA, ATR, a protein kinase is activated. This kinase initiates several complex downstream pathways, which causes a halt in the initiation of new replication origins, prevention of mitosis and replication fork stabilization in order to keep the replication bubble open and DNA polymerase complex attached while the damage is being fixed.

G_2 Phase

G_2 phase, or Gap 2 phase, is the third and final subphase of Interphase in the cell cycle directly preceding mitosis. It follows the successful completion of S phase, during which the cell's DNA is replicated. G_2 phase ends with the onset of prophase, the first phase of mitosis in which the cell's chromatin condenses into chromosomes .

Overview

G_2 phase is a period of rapid cell growth and protein synthesis during which the cell readies itself for mitosis. Curiously, G_2 phase is not a necessary part of the cell cycle, as some cell types (particularly young *Xenopus* embryos and some cancers) proceed directly from DNA replication to mitosis. Though much is known about the genetic network which regulates G2 phase and subsequent entry into mitosis, there is still much to be discovered concerning its significance and regulation, particularly in regards to cancer. One hypothesis is that the growth in G_2 phase is regulated as a method of cell size control. Fission yeast (*Schizosaccharomyces pombe*) has been previously shown to employ such a mechanism, via Cdr2-mediated spatial regulation of Wee1 activity. Though Wee1 is a fairly conserved negative regulator of mitotic entry, no general mechanism of cell size control in G2 has yet been elucidated.

Biochemically, the end of G_2 phase occurs when a threshold level of active cyclin B1/CDK1 complex, also known as Maturation promoting factor (MPF) has been reached. The activity of this complex is tightly regulated during G_2. In particular, the G_2 checkpoint arrests cells in G_2 in response to DNA damage through inhibitory regulation of CDK1.

G_2/M Checkpoint

In vertebrate cells, the G_2/M DNA damage checkpoint consists of an arrest of the cell in G_2 just before mitotic entry in response to genotoxic stress (such as UV radiation, oxidative stress, DNA intercalating agents, etc.) in both a p53-dependent and p53-independent manner. DNA damage signals cause activation of the transcription factor p53. CDK1 is directly inhibited by three transcriptional targets of p53: p21, Gadd45, and 14-3-3σ. Inactive Cyclin B1/CDK1 is sequestered in the nucleus by p21, while active Cyclin B1/CDK1 complexes are sequestered in the cytoplasm by 14-3-3σ. Gadd45 disrupts the binding of Cyclin B1 and CDK1 through direct interaction with CDK1. P53 also transcriptionally represses CDK1.

P53-independent G_2 arrest is mainly affected through the actions of Chk1 kinase. DNA damage is sensed by ATM and ATR (Rad3 and Mec1 in yeast), which then signal to Chk1 and Chk2. Chk1 then mediates the degradation of cdc25A, an activator of CDK1. ATR/ATM also activate p53, indicating that these pathways may act synergistically in regulating G_2 arrest.

Both p53-dependent and p53-independent cell cycle arrest are not specific to G2; these same proteins function upstream in DNA damage checkpoints in G1 and S phase as well. In yeast, which has no p53 homolog, G_2 arrest functions through the p53-independent pathway.

End of G_2/Entry into Mitosis

Mitotic entry is determined by a threshold level of active cyclin B1/CDK1 complex. In vertebrates, there are five cyclin B isoforms (B1, B2, B3, B4, and B5), but specific role of each of these isoforms in regulating mitotic entry is still unclear. It is known that cyclin B1 can compensate for loss of both cyclin B2 (and vice versa in *Drosophila*). Cyclin B1/CDK1 activity is regulated both spatially and temporally during G_2 phase to ensure proper entry into mitosis.

Cyclin B1 transcription begins at the end of S phase after DNA replication. Its promoter contains consensus binding sequences for a number of transcription factors, including p53, p21, Ets, Ap-1, NF-Y, c-Myc, TFE3, and USF. Cyclin B1 accumulates in the cytoplasm throughout G_2, where it binds to and activates CDK1's kinase activity. CDK1 activity is modulated primarily through regulation of its inhibitory phosphorylation sites at Thr14 and Tyr15. Wee1 and Myt1 phosphorylate these two residues, with Wee1 acting on the Tyr15 site and Myt1 acting predominantly on the Thr14 site. However, Myt1 has a separate inhibitory effect on CDK1; it can also sequester CDK1 in the cytoplasm via interaction with Myt1's C-terminal domain. CDK1 is dephosphorylated primarily through the actions of Cdc25, which can dephosphorylate both the Thr14 and Tyr15 residues of CDK1. There are three isoforms of Cdc25 (A, B, and C) in mammalian cells, all of which have been shown to have roles in regulation of G_2 phase.

CDK1, in turn, phosphorylates and modulates the activity of Wee1 and the Cdc25 isoforms A and C. Specifically, CDK1 phosphorylation inhibits Wee1 kinase activity, activates Cdc25C phosphatase activity, and stabilizes Cdc25A. Thus, CDK1 forms a positive feedback loop with Cdc25 and a double negative feedback loop with Wee1 (essentially a net positive feedback loop). These loops encode a hysteretic bistable switch in CDK1 activity relative to Cyclin B1 levels. It is thought that this hysteretic behavior ensures that cells commit to mitosis even if cyclin B1 levels falter.

In mammals, cyclin B1/CDK1 translocation to the nucleus is activated by phosphorylation of five serine sites on cyclin B1's cytoplasmic retention site (CRS): S116, S26, S128, S133, and S147. In Xenopus laevis, Cyclin B1 contains four analogous CRS serine phosphorylation sites (S94, S96, S101, and S113) indicating that this mechanism is highly conserved. Nuclear export is also inactivated by phosphorylation of Cyclin B1's Nuclear export signal (NES). The regulators of these phosphorylation sites are still largely unknown but several factors have been identified, including extracellular signal-regulated kinases (ERKs), PLK1, and CDK1 itself. Upon reaching some threshold level of phosphorylation, translocation of cyclin B1/CDK1 to the nucleus is extremely rapid. Once in the nucleus, Cyclin B1/CDK1 phosphorylates many targets in preparation for mitosis, including Histone H1, nuclear lamins, centrosomal proteins, and Microtubule Associated Proteins (MAPs).

Recently, evidence has emerged suggesting a more important role for cyclin A2/CDK complexes in regulating entry into mitosis. Cyclin A2/CDK2 activity begins in early S phase and increases during G_2. Cdc25B has been shown to dephosphorylate Tyr15 on CDK2 in early-to-mid G_2 in a manner similar to the aforementioned CDK1 mechanism. Downregulation of cyclin A2 in U2OS cells increases Wee1 activity and lowers Plk1 and Cdc25C activity. However, cyclin A2/CDK complexes do not function strictly as activators of cyclin B1/CDK1 in G_2, as CDK2 has been shown to be required for activation of the p53-independent G_2 checkpoint activity, perhaps through a stabilizing phosphorylation on Cdc6. CDK2-/- cells also have aberrantly high levels of Cdc25A. Cyclin A2/CDK1 has also been shown to mediate proteosomal destruction of Cdc25B. These pathways are often deregulated in cancer.

Anaphase

Anaphase is the stage of mitosis when replicated chromosomes are split and the daughter chromatids are moved to opposite poles of the cell. Chromosomes also reach their overall maximum condensation in late anaphase, to help chromosome segregation and the re-formation of the nucleus.

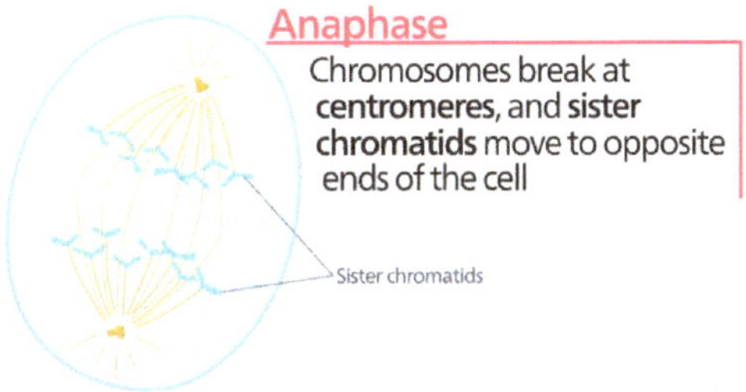

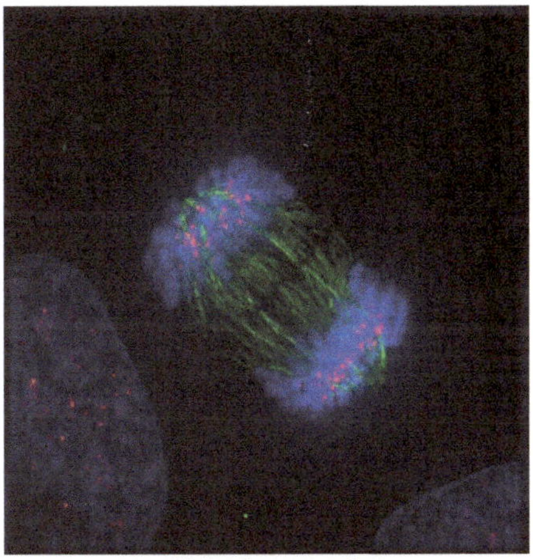

A cell during anaphase.

Anaphase accounts for approximately 1% of the cell cycle's duration. It begins with the regulated triggering of the metaphase-to-anaphase transition. Metaphase ends with the destruction of B cyclin. B cyclin is marked with ubiquitin which flags it for destruction by proteasomes, which is required for the function of metaphase cyclin-dependent kinases (M-Cdks). Anaphase starts when the anaphase promoting complex marks an inhibitory chaperone called securin with ubiquitin for destruction. Securin is a protein which inhibits a protease known as separase. The destruction of securin unleashes separase which then breaks down cohesin, a protein responsible for holding sister chromatids together. The centromeres are split, and the new daughter chromosomes are pulled toward the poles. They take on a V-shape or Y-shape as they are pulled back.

While the chromosomes are drawn to each side of the cell, the non-kinetochore spindle fibers push against each other, in a ratcheting action, that stretches the cell into an oval.

Once anaphase is complete, the cell moves into telophase.

Metaphase

Metaphase is a stage of mitosis in the eukaryotic cell cycle in which chromosomes are at their second-most condensed and coiled stage (they are at their most condensed in anaphase). These chromosomes, carrying genetic information, align in the equator of the cell before being separated into each of the two daughter cells. Metaphase accounts for approximately 4% of the cell cycle's duration. Preceded by events in prometaphase and followed by anaphase, microtubules formed in prophase have already found and attached themselves to kinetochores in metaphase.

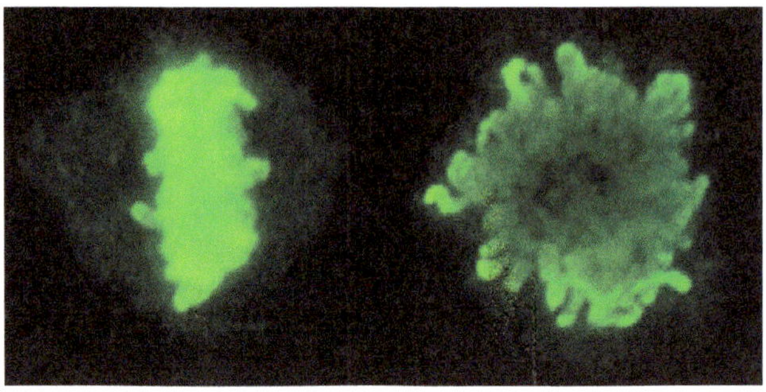

Chromosomes lined up on the metaphase plate. Two views with the metaphase plate rotated 90°.

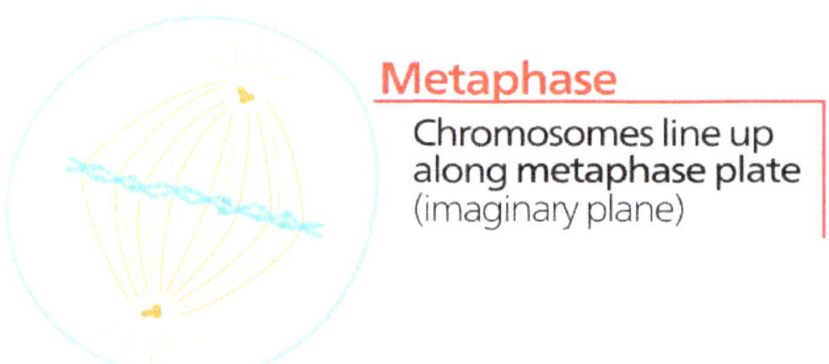

Metaphase in cells (here an animal cell) are characterized by the arrangement of chromosomes at the equatorial plane of the cell

In metaphase, the centromeres of the chromosomes convene themselves on the *metaphase plate* (or *equatorial plate*), an imaginary line that is equidistant from the two centrosome poles. This even alignment is due to the counterbalance of the pulling powers generated by the opposing kinetochore microtubules, analogous to a tug-of-war between two people of equal strength, ending with the destruction of B cyclin. In certain types of cells, chromosomes do not line up at the metaphase plate and instead move back and forth between the poles randomly, only roughly lining up along the middleline. Early events of metaphase can coincide with the later events of prometaphase, as chromosomes with connected kinetochores will start the events of metaphase individually before other chromosomes with unconnected kinetochores that are still lingering in the events of prometaphase.

One of the cell cycle checkpoints occurs during prometaphase and metaphase. Only after all chromosomes have become aligned at the *metaphase plate*, when every kinetochore is properly attached to a bundle of microtubules, does the cell enter anaphase. It is thought that unattached or improperly attached kinetochores generate a signal to prevent premature progression to anaphase, even if most of the kinetochores have been attached and most of the chromosomes have been aligned. Such a signal creates the mitotic spindle checkpoint. This would be accomplished by regulation of the anaphase-promoting complex, securin, and separase.

Metaphase in Cytogenetics and Cancer Studies

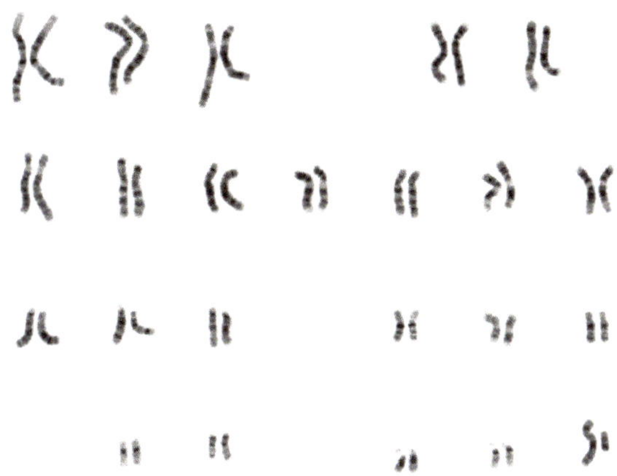

Human metaphase chromosomes (normal male karyotype)

The analysis of metaphase chromosomes is one of the main tools of classical cytogenetics and cancer studies. Chromosomes are condensed (thickened) and highly coiled in metaphase, which makes them most suitable for visual analysis. Metaphase chromosomes make the classical picture of chromosomes (karyotype). For classical cytogenetic analyses, cells are grown in short term culture and arrested in metaphase using mitotic inhibitor. Further they are used for slide preparation and banding (staining) of chromosomes to be visualised under microscope to study structure and number of chromosomes (karyotype). Staining of the slides, often with Giemsa (G banding) or Quinacrine, produces a pattern of in total up to several hundred bands. Normal metaphase spreads are used in methods like FISH and as a hybridization matrix for comparative genomic hybridization (CGH) experiments.

Malignant cells from solid tumors or leukemia samples can also be used for cytogenetic analysis to generate metaphase preparations. Inspection of the stained metaphase chromosomes allows the determination of numerical and structural changes in the tumor cell genome, for example, losses of chromosomal segments or translocations, which may lead to chimeric oncogenes, such as bcr-abl in chronic myelogenous leukemia.

Prophase

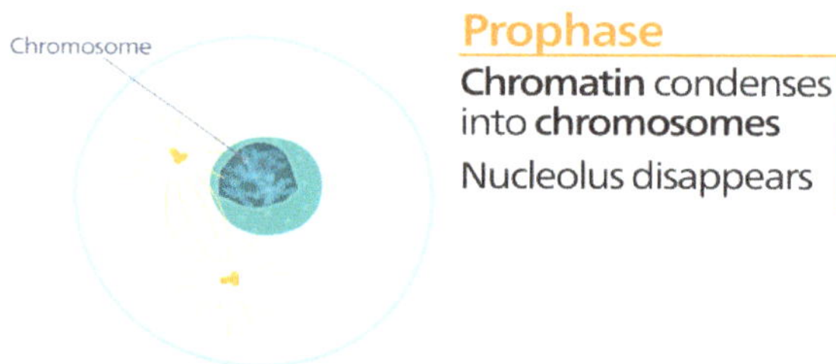

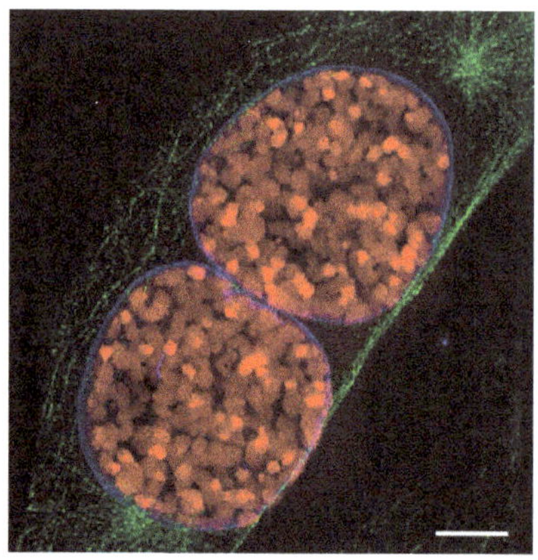

Fluorescence microscope image of two mouse cell nuclei in prophase (scale bar is 5 μm).

Prophase is a stage of mitosis in which the chromatin condenses into double rod-shaped structures called chromosomes in which the chromatin becomes visible. This process, called chromatin condensation, is involved with the condensin complex. Since the genetic material has been replicated in the prior interphase of the cell cycle, there are two identical copies of each chromosome in the cell. Those copies are called sister chromatids and they are attached to each other at a DNA element present on every chromosome called the centromere. Also during prophase, giemsa staining can be applied to elicit G-banding in chromosomes.

An important organelle in mitosis is the centrosome, the microtubule organizing center in metazoans. During prophase, the two centrosomes, which replicate independently of mitosis, have their microtubule-activity increased due to the recruitment of γ-tubulin. The centrosomes will be pushed apart to opposite ends of the cell nucleus by the action of molecular motors acting on the microtubules.

At the end of prophase, the tiny nucleolus within the nucleus dissolves. The nucleolus starts moving towards the nuclear membrane at the beginning of prophase, and its disintegration occurs near the nuclear envelope. Its contents are dispersed in masses.

Basically, the chromatin in the nucleus coils into chromosomes. The centrosomes move to opposite poles of the cell, forming a bridge of spindle fibers. At the end of prophase, the nucleolus disperses.

In prometaphase, the next step of mitosis, the nuclear membrane breaks apart and the chromosomes are captured by the microtubules which are attached to centromeres.

Preprophase

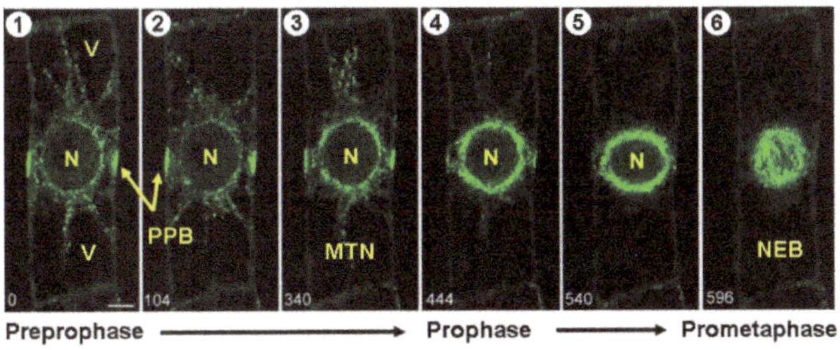

Microtubule dynamics during preprophase and prophase in plant cell mitosis, modified from Donukshe et al. The images follow a tobacco BY-2 cell through the first stages of mitosis (c. 12 minutes). The growing ends of microtubules are shown in green (labeled with green fluorescent protein fused to the microtubule plus end binding protein EB1 of *Arabidopsis thaliana*). N = Nucleus, V = Vacuole, PPB = Preprophase band, MTN = Microtubule nucleation starts at the nuclear envelope, NEB = Nuclear envelope breakdown at the onset of prometaphase.

Preprophase is an additional phase during mitosis in plant cells that does not occur in other eukaryotes such as animals or fungi. It precedes prophase and is characterized by two distinct events:

- The formation of the preprophase band, a dense microtubule ring underneath the plasma membrane.
- The initiation of microtubule nucleation at the nuclear envelope.

Function of Preprophase in the Cell Cycle

Plant cells are fixed with regards to their neighbor cells within the tissues they are growing in. In contrast to animals where certain cells can migrate within the embryo to form new tissues, the seedlings of higher plants grow entirely based on the orientation of cell division and subsequent elongation and differentiation of cells within their cell walls. Therefore, the accurate control of cell division planes and placement of the future cell wall in plant cells is crucial for the correct architecture of plant tissues and organs.

The preprophase stage of somatic plant cell mitosis serves to establish the precise location of the division plane and future cell wall before the cell enters prophase. This is achieved through the formation of a transient microtubule structure, the preprophase band, and a so far unknown mechanism by which the cell is able to "memorize" the position of the preprophase band to guide the new cell wall growing during cytokinesis to the correct location. In gametophyte tissues during the reproductive phase of the plant life cycle, cell division planes may be established without the use of a preprophase band.

In highly vacuolated plant cells, preprophase may be preceded by the formation of a phragmosome. The function of the phragmosome is to suspend the cell nucleus in the center of the cell in preparation for mitosis. If a phragmosome is visible, the preprophase band will appear at its outer edge.

Preprophase Band Formation

At the beginning of preprophase, the cortical microtubules of a plant cell disappear and aggregate into a dense ring underneath the plasma membrane. This preprophase band runs around the equatorial plane of the future mitotic spindle and marks the plane of cell division and future fusion site for the cell plate. It consists of microtubules and microfilaments (actin) and persists into prophase. Spindle formation occurs during prophase with the axis perpendicular to the plane surrounded by the preprophase band.

Microtubule Nucleation

In contrast to animal cells, plant cells do not possess centrosomes to organize their mitotic spindles. Instead, the nuclear envelope acts as a microtubule organizing center (MTOC) for spindle formation during preprophase. The first sign is a clear, actin-free zone appearing around the nuclear envelope. This zone fills with microtubules nucleating on the surface of the nucleus. The preprophase spindle forms by self-assembly of these microtubules in the cytoplasm surrounding the nuclear envelope. It is reinforced through chromosome (kinetochore)-mediated spindle assembly after the nuclear envelope breaks down at the end of prophase.

Transition into Prophase

During progression from preprophase into prophase, the randomly oriented microtubules align parallel along the nuclear surface according to the spindle axis. This structure is called the *prophase spindle*. Triggered by nuclear membrane breakdown at the beginning of prometaphase, the preprophase band disappears and the prophase spindle matures into the metaphase spindle occupying the space of the former nucleus. Experiments with drugs destroying microfilaments indicate that actin may play a role in keeping the cellular "memory" of the position of the division plane after the preprophase band breaks down to direct cytokinesis in telophase.

Prometaphase

In prometaphase, the nuclear membrane breaks apart into numerous "membrane vesicles", and the chromosomes inside form protein structures called kinetochores. Kinetochore microtubules emerging from the centrosomes at the poles (ends) of the spindle reach the chromosomes and attach to the kinetochores, throwing the chromosomes into agitated motion. Other spindle microtubules make contact with microtubules coming from the opposite pole. Forces exerted by protein "motors" associated with spindle microtubules move the chromosomes toward the center of the cell.

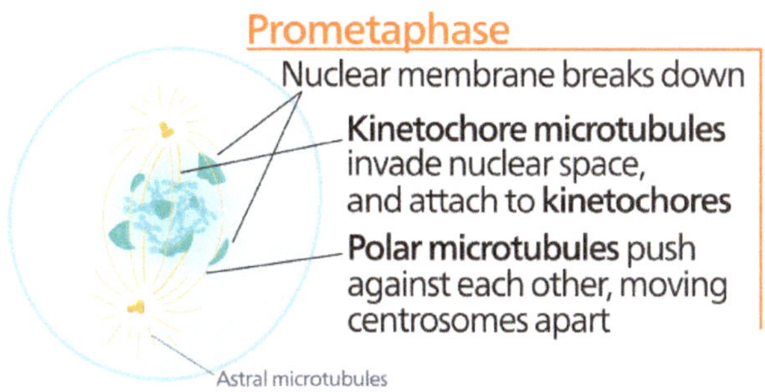

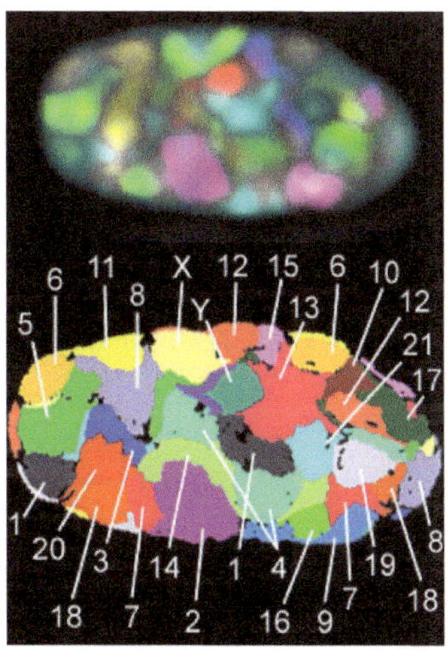

The 23 human chromosome territories during prometaphase in fibroblast cells.

Prometaphase is the phase of mitosis following prophase and preceding metaphase, in eukaryotic somatic cells. Prometaphase is sometimes simply included as part of the end of prophase and early metaphase.

Types of Microtubules

The microtubules are composed of two types, *kinetochore microtubules* and *non-kinetochore microtubules*.

- Kinetochore microtubules begin searching for kinetochores to attach to.
- A number of *non-kinetochore microtubules* or *polar microtubules* find and interact with corresponding nonkinetochore microtubules from the opposite centrosome to form the mitotic spindle.

Transition from Prometaphase to Metaphase

The role of prometaphase is completed when all of the kinetochore microtubules have attached to their kinetochores, upon which metaphase begins. An unattached kinetochore, and thus a non-aligned chromosome, even when most of the other chromosomes have lined up, will trigger the spindle checkpoint signal. This prevents premature progression into anaphase by inhibiting the anaphase-promoting complex until all kinetochores are attached and all the chromosomes aligned.

Early events of metaphase can coincide with the later events of prometaphase, as chromosomes with connected kinetochores will start the events of metaphase individually before other chromosomes with unconnected kinetochores that are still lingering in the events of prometaphase.

Telophase

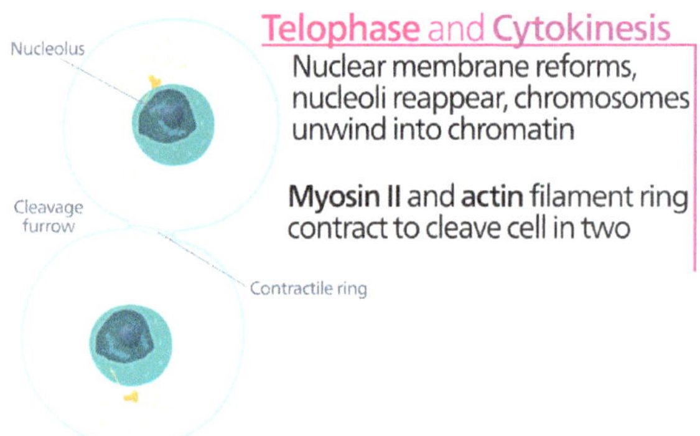

Telophase is the final stage in both meiosis and mitosis in a eukaryotic cell.

During telophase, the effects of prophase and prometaphase (the nuclear membrane and nucleolus disintegrating) are reversed. Two daughter nuclei form in each daughter cell, and phosphatases de-phosphorylate the nuclear lamins at the ends of the cell, forming nuclear envelopes around each nucleus from components of the parent cell's nuclear envelope. Two theories as to how this happens are:

- Vesicle fusion – When fragments of the nuclear membrane fuse to rebuild the nuclear membrane
- Reshaping of the endoplasmic reticulum – where the parts of the endoplasmic reticulum containing absorbed nuclear membrane envelop the nuclear space, reforming a closed membrane.

The chromosomes reach the poles as the nuclear membranes re-form around each set of chromatids, the nucleoli also reappear. The remaining spindle microtubules begin to depolymerize. The chromosomes also unwind back into the expanded chromatin that is present during interphase. Telophase accounts for approximately 2% of the cell cycle's duration.

Reappearance of the nuclear membrane and nucleolus: the telophase

Cytokinesis typically begins before late telophase.

In land plant cells, vesicles derived from the Golgi apparatus move to the middle of the cell along a microtubule scaffold called the phragmoplast. This structure directs packets of cell wall materials which coalesce into a disk-shaped structure called a cell plate. The cell plate grows out centrifugally and eventually develops into a proper cell wall, separating the two nuclei.

References

- Marieb, Elaine (2000), Essentials of human anatomy and physiology, San Francisco: Benjamin Cummings, ISBN 0805349405
- Yael, Avisar; Choi, Jung; DeSaix, Jean; Jurukovski, Vladimir; Wise, Robert; Rye, Connie (2013). Biology. Rice University 6100 Main Street MS-375 Houston, Texas 77005: OpenStax College. p. 281-283. ISBN 978-1-938168-09-3.
- Reece, Jane; Urry, Lisa; Cain, Michael; Wasserman, Steven; Minorsky, Peter; Jackson, Robert (2011). Campbell Biology (10th ed.). Pearson. ISBN 978-0-321-77565-8.
- Zeng, X.; Jiao, M.; Wang, X.; Song, Z.; Hao, S. (2001). "Electron microscopic studies on the Silver-stained Nucleolar Cycle of Physarum Polycephalum" (PDF). Acta Botanica Cinica. 43 (7): 680–5. Retrieved 24 February 2015.
- May, Karen M; Kevin G. Hardwick (2006). "The spindle checkpoint". Journal of Cell Science. 119 (Pt 20): 4139–42. doi:10.1242/jcs.03165. PMID 17038540. Retrieved 9 December 2012.

7

Cells in Various Life Cycles

Cells can be divided based on the function and genetic variation they display. This chapter explores the various types of cells like germ cells, somatic cells, zygote, gamete and phragmoplast. The content provides information on the differences between the various types.

Germ Cell

A germ cell is any biological cell that gives rise to the gametes of an organism that reproduces sexually. In many animals, the germ cells originate in the primitive streak and migrate via the gut of an embryo to the developing gonads. There, they undergo meiosis, followed by cellular differentiation into mature gametes, either eggs or sperm. Unlike animals, plants do not have germ cells set aside from in early development. Instead, germ cells can come from somatic cells in the adult (such as the floral meristem of flowering plants).

Introduction

Multicellular eukaryotes are made of two fundamental cell types. Germ cells produce gametes and are the only cells that can undergo meiosis as well as mitosis. These cells are sometimes said to be immortal because they are the link between generations. Somatic cells are all the other cells that form the building blocks of the body and they only divide by mitosis. The lineage of germ cells is called germ line. Germ cell specification begins during cleavage in many animals or in the epiblast during gastrulation in birds and mammals. After transport, involving passive movements and active migration, germ cells arrive at the developing gonads. In humans, sexual differentiation starts approximately 6 weeks after conception. The end-products of the germ cell cycle are the egg or sperm.

Under special conditions *in vitro* germ cells can acquire properties similar to those of embryonic stem cells (ES). The underlying mechanism of that change is still unknown. These changed cells are then called embryonic germ cells (EG). Both EG and ES are pluripotent in vitro, but only ES has proven pluripotency in vivo. Recent studies have demonstrated that it is possible to give rise to primordial germ cells from ES.

Specification

There are two mechanisms to establish the germ cell lineage in the embryo. The first way is called preformistic and involves that the cells destined to become germ cells inherit the specific germ cell determinants present in the germ plasm (specific area of the cytoplasm) of the egg (ovum). The unfertilized egg of most animals is asymmetrical: different regions of the cytoplasm contain different amounts of mRNA and proteins. By this, germ cells obtained by the first divisions of the fertilized egg are characterized by specific molecules of a particular region of the egg cytoplasm.

The second way is found in birds and mammals, where germ cells are not specified by such determinants but by signals controlled by zygotic genes. In mammals, a few cells of the early embryo are induced by signals of neighboring cells to become primordial germ cells. Mammalian eggs are somewhat symmetrical and after the first divisions of the fertilized egg, the produced cells are all totipotent. This means that they can differentiate in any cell type in the body and thus germ cells. Specification of primordial germ cells in the laboratory mouse is initiated by high levels of bone morphogenetic protein (BMP) signaling, which activates expression of the transcription factors Blimp-1/ Prdm1 and Prdm14.

Migration

Primordial germ cells, germ cells that still have to reach the gonads, also known as PGCs, precursor germ cells or gonocytes, divide repeatedly on their migratory route through the gut and into the developing gonads.

Invertebrates

In the model organism *Drosophila*, pole cells passively move from the posterior end of the embryo to the posterior midgut because of the infolding of the blastoderm. Then they actively move through the gut into the mesoderm. Endodermal cells differentiate and together with Wunen proteins they induce the migration through the gut. Wunen proteins are chemorepellents that lead the germ cells away from the endoderm and into the mesoderm. After splitting into two populations, the germ cells continue migrating laterally and in parallel until they reach the gonads. Columbus proteins, chemoattractants, stimulate the migration in the gonadal mesoderm.

Vertebrates

In the *Xenopus* egg, the germ cell determinants are found in the most vegetal blastomeres. These presumptive PGCs are brought to the endoderm of the blastocoel by gastrulation. They are determined as germ cells when gastrulation is completed. Migration from the hindgut along the gut and across the dorsal mesentery then takes place. The germ cells split into two populations and move to the paired gonadal ridges. Migration starts with 3-4 cells that undergo three rounds of cell division so that about 30 PGCs arrive at the gonads. On the migratory path of the PGCs, the orientation of underlying cells and their secreted molecules such as fibronectin play an important role.

Mammals have a migratory path comparable to that in *Xenopus*. Migration begins with 50 gonocytes and about 5,000 PGCs arrive at the gonads. Proliferation occurs also during migration and lasts for 3–4 weeks in humans.

PGCs come from the epiblast and migrate subsequently into the mesoderm, the endoderm and the posterior of the yolk sac. Migration then takes place from the hindgut along the gut and across the dorsal mesentery to reach the gonads (4.5 weeks in human beings). Fibronectin maps here also a polarized network together with other molecules. The somatic cells on the path of germ cells provide them attractive, repulsive, and survival signals. But germ cells also send signals to each other.

In reptiles and birds, germ cells use another path. PGCs come from the epiblast and move to the hypoblast to form the germinal crescent (anterior extraembryonic structure). The gonocytes then squeeze into blood vessels and use the circulatory system for transport. They squeeze out of the vessels when they are at height of the gonadal ridges. Cell adhesion on the endothelium of the blood vessels and molecules such as chemoattractants are probably involved in helping PGCs migrate.

The Sry Gene of the Y Chromosome

The *SRY* (Sex-determining Region of the Y chromosome) directs male development in mammals by inducing the somatic cells of the gonadal ridge to develop into a testis, rather than an ovary. *Sry* is expressed in a small group of somatic cells of the gonads and influences these cells to become Sertoli cells (supporting cells in testis). Sertoli cells are responsible for sexual development along a male pathway in many ways. One of these ways involves stimulation of the arriving primordial cells to differentiate into sperm. In the absence of the *Sry* gene, primordial germ cells differentiate into eggs. Removing genital ridges before they start to develop into testes or ovaries results in the development of a female, independent of the carried sex chromosome.

Gametogenesis

Gametogenesis, the development of diploid germ cells into either haploid eggs or sperm, (respectively oogenesis and spermatogenesis) is different for each species but the general stages are similar. Oogenesis and spermatogenesis have many features in common, they both involve:

- Meiosis
- Extensive morphological differentiation
- Incapacity of surviving for very long if fertilization does not occur

Despite their homologies they also have major differences:

- Spermatogenesis has equivalent meiotic divisions resulting in four equivalent spermatids while oogenic meiosis is asymmetrical: only one egg is formed together with three polar bodies.

- Different timing of maturation: oogenic meiosis is interrupted at one or more stages (for a long time) while spermatogenic meiosis is rapid and uninterrupted.

Oogenesis

After migration primordial germ cells will become oogonia in the forming gonad (ovary). The oogonia proliferate extensively by mitotic divisions, up to 5-7 million cells in humans. But then many of these oogonia die and about 50,000 remain. These cells differentiate into primary oocytes. In week 11-12 *post coitus* the first meiotic division begins (before birth for most mammals) and remains arrested in prophase I from a few days to many years depending on the species. It is in this period or in some cases at the beginning of sexual maturity that the primary oocytes secrete proteins to form a coat called zona pellucida and they also produce cortical granules containing enzymes and proteins needed for fertilization. Meiosis stands by because of the follicular granulosa cells that send inhibitory signals through gap junctions and the zona pellucida. Sexual maturation is the beginning of periodic ovulation. Ovulation is the regular release of one oocyte from the ovary into the reproductive tract and is preceded by follicular growth. A few follicle cells are stimulated to grow but only one oocyte is ovulated. A primordial follicle consists of an epithelial layer of follicular granulosa cells enclosing an oocyte. The pituitary gland secrete follicle-stimulating hormones (FSHs) that stimulate follicular growth and oocyte maturation. The thecal cells around each follicle secrete estrogen. This hormone stimulates the production of FSH receptors on the follicular granulosa cells and has at the same time a negative feedback on FSH secretion. This results in a competition between the follicles and only the follicle with the most FSH receptors survives and is ovulated. Meiotic division I goes on in the ovulated oocyte stimulated by luteinizing hormones (LHs) produced by the pituitary gland. FSH and LH block the gap junctions between follicle cells and the oocyte therefore inhibiting communication between them. Most follicular granulosa cells stay around the oocyte and so form the cumulus layer. Large non-mammalian oocytes accumulate egg yolk, glycogen, lipids, ribosomes, and the mRNA needed for protein synthesis during early embryonic growth. These intensive RNA biosynthese are mirrored in the structure of the chromosomes, which decondense and form lateral loops giving them a lampbrush appearance. Oocyte maturation is the following phase of oocyte development. It occurs at sexual maturity when hormones stimulate the oocyte to complete meiotic division I. The meiotic division I produces 2 cells differing in size: a small polar body and a large secondary oocyte. The secondary oocyte undergoes meiotic division II and that results in the formation of a second small polar body and a large mature egg, both being haploid cells. The polar bodies degenerate. Oocyte maturation stands by at metaphase II in most vertebrates. During ovulation, the arrested secondary oocyte leaves the ovary and matures rapidly into an egg ready for fertilization. Fertilization will cause the egg to complete meiosis II. In human females there is proliferation of the oogonia in the fetus, meiosis starts then before birth and stands by at meiotic division I up to 50 years, ovulation begins at puberty.

Egg Growth

A 10 - 20 µm large somatic cell generally needs 24 hours to double its mass for mitosis. By this way it would take a very long time for that cell to reach the size of a mammalian egg with a diameter of 100 µm (some insects have eggs of about 1,000 µm or greater). Eggs have therefore special mechanisms to grow to their large size. One of these mechanisms is to have extra copies of genes: meiotic division I is paused so that the oocyte grows while it contains two diploid chromosome sets. Some species produce many extra copies of genes, such as amphibians, which may have up to 1 or 2 million copies. A complementary mechanism is partly dependent on syntheses of other cells. In amphibians, birds, and insects, yolk is made by the liver (or its equivalent) and secreted into the blood. Neighboring accessory cells in the ovary can also provide nutritive help of two types. In some invertebrates some oogonia become nurse cells. These cells are connected by cytoplasmic bridges with oocytes. The nurse cells of insects provide oocytes macromolecules such as proteins and mRNA. Follicular granulosa cells are the second type of accessory cells in the ovary in both invertebrates and vertebrates. They form a layer around the oocyte and nourish them with small molecules, no macromolecules, but eventually their smaller precursor molecules, by gap junctions.

Spermatogenesis

Mammalian spermatogenesis is representative for most animals. In human males, spermatogenesis begins at puberty in seminiferous tubules in the testicles and go on continuously. Spermatogonia are immature germ cells. They proliferate continuously by mitotic divisions around the outer edge of the seminiferous tubules, next to the basal lamina. Some of these cells stop proliferation and differentiate into primary spermatocytes. After they proceed through the first meiotic division, two secondary spermatocytes are produced. The two secondary spermatocytes undergo the second meiotic division to form four haploid spermatids. These spermatids differentiate morphologically into sperm by nuclear condensation, ejection of the cytoplasm and formation of the acrosome and flagellum.

The developing male germ cells do not complete cytokinesis during spermatogenesis. Consequently, cytoplasmic bridges assure connection between the clones of differentiating daughter cells to form a syncytium. In this way the haploid cells are supplied with all the products of a complete diploid genome. Sperm that carry a Y chromosome, for example, is supplied with essential molecules that are encoded by genes on the X chromosome.

Diseases

Germ cell tumor is a rare cancer that can affect people at all ages. 2.4 children out of

1 million suffer the disease, and it counts for 4% of all cancers in children and adolescents younger than 20 years old.

Germ cell tumors are generally located in the gonads but can also appear in the abdomen, pelvis, mediastinum, or brain. Germ cells migrating to the gonads may not reach that intended destination and a tumor can grow wherever they end up, but the exact cause is still unknown. These tumors can be benign or malignant.

Induced Differentiation

Inducing differentiation of certain cells to germ cells has many applications. One implication of induced differentiation is that it may allow for the eradication of male and female factor infertility. Furthermore, it would allow same-sex couples to have biological children if sperm could be produced from female cells or if eggs could be produced from male cells. Efforts to create sperm and eggs from skin and embryonic stem cells were pioneered by Hayashi and Saitou's research group at Kyoto University. These researchers produced primordial germ cell-like cells (PGCs) from embryonic stem cells (ESCs) and skin cells in vitro.

Hayashi and Saitou's group was able to promote the differentiation of embryonic stem cells into PGCs with the use of precise timing and bone morphogenetic protein 4 (Bmp4). The PGCs were then placed in the testes of mice that were initially unable to produce their own sperm cells. As a result, the mice were able to produce sperm cells. Upon succeeding with embryonic stem cells, the group was able to successfully promote the differentiation of induced pluripotent stem cells (iPSCs) into PGCs. These primordial germ cell-like cells were then used to create spermatozoa and oocytes.

Efforts for human cells are less advanced due to the fact that the PGCs formed by these experiments are not always viable. In fact Hayashi and Saitou's method is only one third as effective as current in vitro fertilization methods, and the produced PGCs are not always functional. Furthermore, not only are the induced PGCs not as effective as naturally occurring PGCs, but they are also less effective at erasing their epigenetic markers when they differentiate from iPSCs or ESCs to PGCs.

There are also other applications of induced differentiation of germ cells. Another study showed that culture of human embryonic stem cells in mitotically inactivated porcine ovarian fibroblasts (POF) causes differentiation into germ cells, as evidenced by gene expression analysis.

Somatic Cell

A somatic or vegetal cell is any biological cell forming the body of an organism; that is, in a multicellular organism, any cell other than a gamete, germ cell, gametocyte or undifferentiated stem cell.

In contrast, gametes are cells that fuse during sexual reproduction, germ cells are cells that give rise to gametes, and stem cells are cells that can divide through mitosis and differentiate into diverse specialized cell types. For example, in mammals, somatic cells make up all the internal organs, skin, bones, blood and connective tissue, while mammalian germ cells give rise to spermatozoa and ova which fuse during fertilization to produce a cell called a zygote, which divides and differentiates into the cells of an embryo. There are approximately 220 types of somatic cells in the human body.

The word "somatic" is derived from the Greek word *sōma*, meaning "body".

Evolution

As multicellularity evolved many times, sterile somatic cells did too. The evolution of an immortal germline producing specialized somatic cells involved the emergence of mortality, and can be viewed in its simplest version in volvocine algae. Those species with a separation between sterile somatic cells and a germ line are called Weismannists. However, Weismannist development is relatively rare (e.g., vertebrates, arthropods, *Volvox*), as great part of species have the capacity for somatic embryogenesis (e.g., land plants, most algae, many invertebrates).

Genetics and Chromosome Content

Like all cells, somatic cells contain DNA arranged in chromosomes. If a somatic cell contains chromosomes arranged in pairs, it is called diploid and the organism is called a diploid organism. (The gametes of diploid organisms contain only single unpaired chromosomes and are called haploid.) Each pair of chromosomes comprises one chromosome inherited from the father and one inherited from the mother. For example, in humans, somatic cells contain 46 chromosomes organized into 23 pairs. By contrast, gametes of diploid organisms contain only half as many chromosomes. In humans, this is 23 unpaired chromosomes. When two gametes (i.e. a spermatozoon and an ovum) meet during conception, they fuse together, creating a zygote. Due to the fusion of the two gametes, a human zygote contains 46 chromosomes (i.e. 23 pairs).

However, a large number of species have the chromosomes in their somatic cells arranged in fours ("tetraploid") or even sixes ("hexaploid"). Thus, they can have diploid or even triploid germline cells. An example of this is the modern cultivated species of wheat, *Triticum aestivum L.*, a hexaploid species whose somatic cells contain six copies of every chromatid.

Cloning

In recent years, the technique of cloning whole organisms has been developed in mammals, allowing almost identical genetic clones of an animal to be produced. One method of doing this is called "somatic cell nuclear transfer" and involves removing the nucleus from a somatic cell, usually a skin cell. This nucleus contains all of the genetic information needed to produce the organism it was removed from. This nucleus is then injected into an ovum of the same species which has had its own genetic material removed. The ovum now no longer needs to be fertilized, because it contains the correct amount of genetic material (a diploid number of chromosomes). In theory, the ovum can be implanted into the uterus of a same-species animal and allowed to develop. The resulting animal will be a nearly genetically identical clone to the animal from which the nucleus was taken. The only difference is caused by any mitochondrial DNA that is retained in the ovum, which is different from the cell that donated the nucleus. In practice, this technique has so far been problematic, although there have been a few high profile successes, such as Dolly the Sheep and, more recently, Snuppy, the first cloned dog. Somatic cells have also been collected in the practice of Cryoconservation of animal genetic resources as a means of conserving animal genetic material, including to clone livestock.

Somatic Cell Modifications

Development of Biotechnology has allowed for the genetic manipulation of somatic cells.This biotechnology deals with some ethical controversy in Human genetic engineering.

Zygote

A zygote is a eukaryotic cell formed by a fertilization event between two gametes. The zygote's genome is a combination of the DNA in each gamete, and contains all of the genetic information necessary to form a new individual. In multicellular organisms, the zygote is the earliest developmental stage. In single-celled organisms, the zygote can divide asexually by mitosis to produce identical offspring.

Oscar Hertwig and Richard Hertwig made some of the first discoveries on animal Zygote Formation.

Fungi

In fungi, the sexual fusion of haploid cells is called karyogamy. The result of karyogamy is a diploid cell called a zygote or zygospore. This cell may then enter meiosis or mitosis depending on the life cycle of the species.

Plants

In plants, the zygote may be polyploid if fertilization occurs between meiotically unreduced gametes.

In land plants, the zygote is formed within a chamber called the archegonium. In seedless plants, the archegonium is usually flask-shaped, with a long hollow neck through which the sperm cell enters. As the zygote divides and grows, it does so inside the archegonium.

Humans

In human fertilization, two 1n haploid cells—an ovum (female gamete) and a sperm cell (male gamete)—combine to form a single 2n diploid cell called the zygote. DNA is then replicated in the two separate pronuclei derived from the sperm and ovum, making the zygote's chromosome number temporarily 4n diploid. After approximately 30 hours, fusion of the pronuclei and subsequent mitotic division produce two 2n diploid daughter cells called blastomeres.

Between the stages of fertilization and implantation, the developing human is called the *preimplantation conceptus* or the proembryo. It is not correct to call the conceptus an *embryo*, because it will later differentiate into both intraembryonic and extraembryonic tissues, and can even split to produce multiple embryos (identical twins).

After fertilization, the conceptus travels down the oviduct towards the uterus while continuing to divide mitotically without actually increasing in size, in a process called cleavage. After four divisions, the conceptus consists of 16 blastomeres, and it is known as the morula. Through the processes of compaction, cell division, and blastulation, the conceptus takes the form of the blastocyst by the fifth day of development, just as it approaches the site of implantation. When the blastocyst hatches from the zona pellucida, it can implant in the endometrial lining of the uterus and begin the embryonic stage of development.

The human zygote has been genetically edited in experiments designed to cure inherited diseases.

In Other Species

A Chlamydomonas zygote that contains chloroplast DNA (cpDNA) from both parents, such cells generally are rare since normally cpDNA is inherited uniparental from the mt+ mating type parent.These rare biparental zygotes allowed mapping of chloroplast genes by recombination.

In Protozoa

In the amoeba, reproduction occurs by cell division of the parent cell: first the nucleus of the parent divides into two and then the cell membrane also cleaves, becoming two "daughter" Amoebae.

Gamete

A gamete is a cell that fuses with another cell during fertilization (conception) in organisms that sexually reproduce. In species that produce two morphologically distinct types of gametes, and in which each individual produces only one type, a female is any individual that produces the larger type of gamete—called an ovum (or egg)—and a male produces the smaller tadpole-like type—called a sperm. This is an example of anisogamy or heterogamy, the condition in which females and males produce gametes of different sizes (this is the case in humans; the human ovum has approximately 100,000 times the volume of a single human sperm cell). In contrast, isogamy is the state of gametes from both sexes being the same size and shape, and given arbitrary designators for mating type. The name gamete was introduced by the Austrian biologist Gregor Mendel. Gametes carry half the genetic information of an individual, one ploidy of each type, and are created through meiosis.

Dissimilarity

In contrast to a gamete, the diploid somatic cells of an individual contain one copy of the chromosome set from the sperm and one copy of the chromosome set from the egg cell; that is, the cells of the offspring have genes expressing characteristics of both the *father* and the *mother*. A gamete's chromosomes are not exact duplicates of either of the sets of chromosomes carried in the diploid chromosomes, and often undergo random mutations resulting in modified DNA (and subsequently, new proteins and phenotypes).

Gender Determination in Humans and Birds

In humans, a normal ovum can carry only an X chromosome (of the X and Y chromosomes), whereas a sperm may carry either an X or a Y (a non-normal ovum can end up carrying two or no X chromosomes, as a result of a mistake at either of the two stages of meiosis, while a non-normal sperm cell can end up carrying either no sex-defining chromosomes, an XY pair, an XX pair as a result of the forementioned reason); ergo the male sperm can play a role in determining the gender of any resulting zygote, if the zygote has two X chromosomes it may develop into a female, if it has an X and a Y chromosome, it may develop into a male. For birds, the female ovum determines the sex of the offspring, through the ZW sex-determination system.

Artificial Gametes

Artificial gametes, also known as In vitro derived gametes (IVD), stem cell-derived gametes (SCDGs), and In vitro generated gametes (IVG), are gametes derived from stem cells. Research shows that artificial gametes may be a reproductive technique for same-sex male couples, although a surrogate mother would still be required for the gestation period. Women who have passed menopause may be able to produce eggs and bear genetically related children with artificial gametes. Robert Sparrow wrote, in the Journal of Medical Ethics, that embryos derived from artificial gametes could be used to derive new gametes and this process could be repeated to create multiple human generations in the laboratory. This technique could be used to create cell lines for medical applications and for studying the heredity of genetic disorders. Additionally, this technique could be used for human enhancement by selectively breeding for a desired genome or by using recombinant DNA technology to create enhancements that have not arisen in nature.

Plants

Plants which reproduce sexually also have gametes. However, since plants have an alternation of diploid and haploid generations some differences exist. In flowering plants the flowers use meiosis to produce a haploid generation which produce gametes through mitosis. The female haploid is called the ovule and is produced by the ovary of the flower. When mature the haploid ovule produces the female gamete which are ready for fertilization. The male haploid is pollen and is produced by the anther, when pollen lands on a mature stigma of a flower it grows a pollen tube down into the flower. The haploid pollen then produces sperm by mitosis and releases them for fertilization.

Phragmoplast

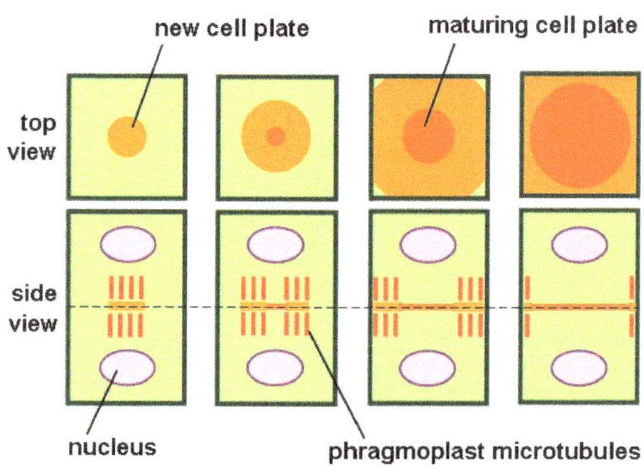

Phragmoplast and cell plate formation in a plant cell during cytokinesis. Left side: Phragmoplast forms and cell plate starts to assemble in the center of the cell. Towards the right: Phragmoplast enlarges in a donut-shape towards the outside of the cell, leaving behind mature cell plate in the center. The cell plate will transform into the new cell wall once cytokinesis is complete.

The phragmoplast is a plant cell specific structure that forms during late cytokinesis. It serves as a scaffold for cell plate assembly and subsequent formation of a new cell wall separating the two daughter cells.

The phragmoplast is a complex assembly of microtubules (MTs), microfilaments (MFs), and endoplasmic reticulum (ER) elements, that assemble in two opposing sets perpendicular to the plane of the future cell plate during anaphase and telophase. It is initially barrel-shaped and forms from the mitotic spindle between the two daughter nuclei while nuclear envelopes reassemble around them. The cell plate initially forms as a disc between the two halves of the phragmoplast structure. While new cell plate material is added to the edges of the growing plate, the phragmoplast microtubules disappear in the center and regenerate at the edges of the growing cell plate. The two structures grow outwards until they reach the outer wall of the dividing cell. If a phragmosome was present in the cell, the phragmoplast and cell plate will grow through the space occupied by the phragmosome. They will reach the parent cell wall exactly at the position formerly occupied by the preprophase band.

The microtubules and actin filaments within the phragmoplast serve to guide vesicles with cell wall material to the growing cell plate. Actin filaments are also possibly involved in guiding the phragmoplast to the site of the former preprophase band location at the parent cell wall. While the cell plate is growing, segments of smooth endoplasmic reticulum are trapped within it, later forming the plasmodesmata connecting the two daughter cells.

The phragmoplast can only be observed in Embryophytes, that is the bryophytes and vascular plants and a few advanced green algae, specifically *Coleochaete* in the Division Charophyta. Some algae use another type of microtubule array, a phycoplast, during cytokinesis.

References

- Campbell, Neil A.; Reece, Jane B.; Urry, Lisa A.; Cain, Michael L.; Wasserman, Steven A.; Minorsky, Peter V.; Jackson, Robert B. (2009). Biology (9th ed.). p. 229. ISBN 978-0-8053-6844-4.
- Jay Phelan (30 April 2009). What Is Life?: A Guide to Biology W/Prep-U. Macmillan. pp. 237–. ISBN 978-1-4292-2318-8. Retrieved 8 October 2010.
- P.H. Raven, R.F. Evert, S.E. Eichhorn (2005): Biology of Plants, 7th Edition, W.H. Freeman and Company Publishers, New York, ISBN 0-7167-1007-2
- Newson, A J; Smajdor, A C (2005). "Artificial gametes: new paths to parenthood?" (PDF). Journal of Medical Ethics. 31: 184–186. doi:10.1136/jme.2003.004986. Retrieved 26 February 2015.
- Sparrow, Robert (4 April 2013). "In vitro eugenics". Journal of Medical Ethics. doi:10.1136/medethics-2012-101200. Retrieved 8 March 2015.

8

Homologous Chromosome: An Overview

Homologous chromosomes are chromosomes that align perfectly with one another as they share the same genes and loci. This chapter provides the reader with an overview on homologous chromosomes, sister chromatids, the process of cohesin etc.

Homologous Chromosome

A couple of homologous chromosomes, or homologs, are a set of one maternal and one paternal chromosomes that pair up with each other inside a cell during meiosis. These copies have the same genes in the same loci where they provide points along each chromosome which enable a pair of chromosomes to align correctly with each other before separating during meiosis. This is the basis for Mendelian inheritance which characterizes inheritance patterns of genetic material from an organism to its offspring parent developmental cell at the given time and area.

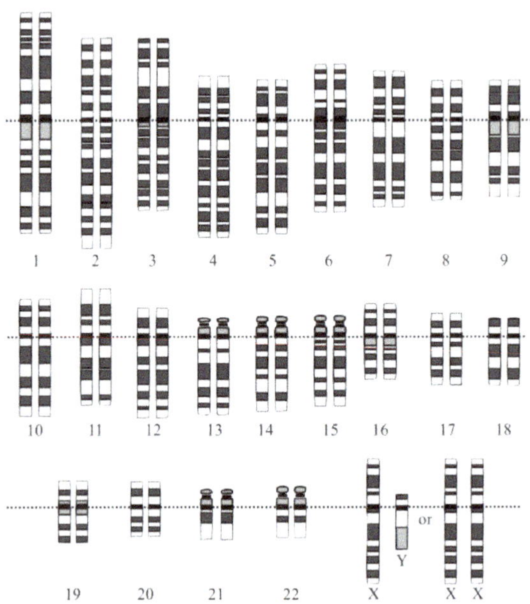

As this karyotype displays, a diploid human cell contains 23 pairs of homologous chromosomes. The cell has two sets of each chromosome; one of the pair is derived from the mother and the other from the father. The maternal and paternal chromosomes in a homologous pair have the same genes at the same loci, but possibly different alleles.

Overview

Chromosomes are linear arrangements of condensed deoxyribonucleic acid (DNA) and histone proteins, which form a complex called chromatin. Homologous chromosomes are made up of chromosome pairs of approximately the same length, centromere position, and staining pattern, for genes with the same corresponding loci. One homologous chromosome is inherited from the organism's mother; the other is inherited from the organism's father. After mitosis occurs within the daughter cells, they have the correct number of genes which are a mix of the two parents' genes. In diploid (2n) organisms, the genome is composed of one set of each homologous chromosome pair, as compared to tetraploid organisms which may have two sets of each homologous chromosome pair. The alleles on the homologous chromosomes may be different, resulting in different phenotypes of the same genes. This mixing of maternal and paternal traits is enhanced by crossing over during meiosis, wherein lengths of chromosomal arms and the DNA they contain within a homologous chromosome pair are exchanged with one another.

History

Early in the 1900s William Bateson and Reginald Punnett were studying genetic inheritance and they noted that some combinations of alleles appeared more frequently than others. That data and information was further explored by Thomas Morgan. Using test cross experiments, he revealed that, for a single parent, the alleles of genes near to one another along the length of the chromosome move together. Using this logic he concluded that the two genes he was studying were located on homologous chromosomes. Later on during the 1930s Harriet Creighton and Barbara McClintock were studying meiosis in corn cells and examining gene loci on corn chromosomes. Creighton and McClintock discovered that the new allele combinations present in the offspring and the event of crossing over were directly related. This proved intrachromosomal genetic recombination.

Structure

Homologous chromosomes are chromosomes which contain the same genes in the same order along their chromosomal arms. There are two main properties of homologous chromosomes: the length of chromosomal arms and the placement of the centromere

The actual length of the arm, in accordance with the gene locations, is critically important for proper alignment. Centromere placement can be characterized by four main arrangements, consisting of being either metacentric, submetacentric, telocentric, or acrocentric. Both of these properties are the main factors for creating structural homology between chromosomes. Therefore, when two chromosomes of the exact structure exist, they are able to pair together to form homologous chromosomes.

Since homologous chromosomes are not identical and do not originate from the same organism, they are different from sister chromatids. Sister chromatids result after DNA replication has occurred, and thus are identical, side-by-side duplicates of each other.

In Humans

Humans have a total of 46 chromosomes, but there are only 22 pairs of homologous autosomal chromosomes. The additional 23rd pair is the sex chromosomes, X and Y. If this pair is made up of an X and Y chromosome, then the pair is not truly homologous because their size and gene content differ greatly. The 22 pairs of homologous chromosomes contain the same genes but code for different traits in their allelic forms since one was inherited from the mother and one from the father. So humans have two homologous chromosome sets in each cell, meaning humans are diploid organisms.

Functions

Homologous chromosomes are important in the processes of meiosis and mitosis. They allow for the recombination and random segregation of genetic material from the mother and father into new cells.

In Meiosis

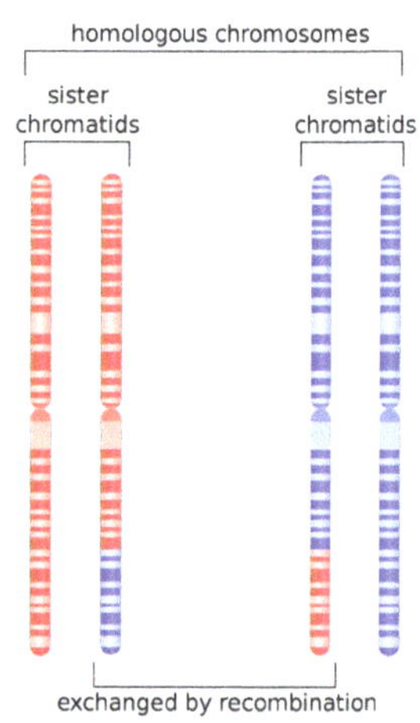

During the process of meiosis, homologous chromosomes can recombine and produce new combinations of genes in the daughter cells.

Meiosis is a round of two cell divisions that results in four haploid daughter cells that each contain half the number of chromosomes as the parent cell. It reduces the chromosome number in a germ cell by half by first separating the homologous chromosomes in meiosis I and then the sister chromatids in meiosis II. The process of meiosis I is generally longer than meiosis II because it takes more time for the chromatin to replicate and for the homologous chromosomes to be properly oriented and segregated by the processes of pairing and synapsis in meiosis I. During meiosis, genetic recombination (by random segregation) and crossing over produces daughter cells that each contain different combinations of maternally and paternally coded genes. This recombination of genes allows for the introduction of new allele pairings and genetic variation. Genetic variation among organisms helps make a population more stable by providing a wider range of genetic traits for natural selection to act on.

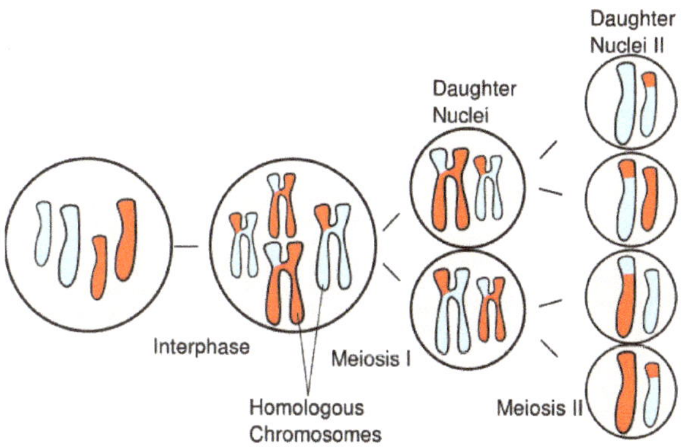

Sorting of homologous chromosomes during meiosis.

Prophase I

In prophase I of meiosis I, each chromosome is aligned with its homologous partner and pairs completely. In prophase I, the DNA has already undergone replication so each chromosome consists of two identical chromatids connected by a common centromere. During the zygotene stage of prophase I, the homologous chromosomes pair up with each other. This pairing occurs by a synapsis process where the synaptonemal complex - a protein scaffold - is assembled and joins the homologous chromosomes along their lengths. Cohesin crosslinking occurs between the homologous chromosomes and helps them resist being pulled apart until anaphase. Genetic crossing over occurs during the pachytene stage of prophase I. In this process, genes are exchanged by the breaking and union of homologous portions of the chromosomes' lengths. Structures called chiasmata are the site of the exchange. Chiasmata physically link the homologous chromosomes once crossing over occurs and throughout the process of chromosomal segregation during meiosis. At the diplotene stage of prophase I the synaptonemal complex disassembles before which will allow the homologous chromosomes to separate, while the sister chromatids stay associated by their centromeres.

Metaphase I

In metaphase I of meiosis I, the pairs of homologous chromosomes, also known as bivalents or tetrads, line up in a random order along the metaphase plate. The random orientation is another way for cells to introduce genetic variation. Meiotic spindles emanating from opposite spindle poles attach to each of the homologs (each pair of sister chromatids) at the kinetochore.

Anaphase I

In anaphase I of meiosis I the homologous chromosomes are pulled apart from each other. The homologs are cleaved by the enzyme separase to release the cohesin that held the homologous chromosome arms together. This allows the chiasmata to release and the homologs to move to opposite poles of the cell. The homologous chromosomes are now randomly segregated into two daughter cells that will undergo meiosis II to produce four haploid daughter germ cells.

Meiosis II

After the tetrads of homologous chromosomes are separated in meiosis I, the sister chromatids from each pair are separated. The two haploid(because the chromosome no. has reduced to half. Earlier two sets of chromosomes were present, but now each set exists in two different daughter cells that have arisen from the single diploid parent cell by meiosis I) daughter cells resulting from meiosis I undergo another cell division in meiosis II but without another round of chromosomal replication. The sister chromatids in the two daughter cells are pulled apart during anaphase II by nuclear spindle fibers, resulting in four haploid daughter cells.

In Mitosis

Homologous chromosomes do not function the same in mitosis as they do in meiosis. Prior to every single mitotic division a cell undergoes, the chromosomes in the parent cell replicate themselves. The homologous chromosomes within the cell will not pair up and undergo genetic recombination with each other. Instead, the replicants, or sister chromatids, will line up along the metaphase plate and then separate in the same way as meiosis II - by being pulled apart at their centromeres by nuclear mitotic spindles. If any crossing over does occur between sister chromatids during mitosis, it does not produce any recombinants.

In Somatic Cells

Homologous pairing in most contexts will refer to germline cells, however also takes place in somatic cells. For example, in humans, somatic cells have very tightly regulated homologous pairing (separated into chromosomal territories, and pairing at specific loci under control of developmental signalling). Other species however (notably *Drosophila*) exhibit homologous pairing much more frequently. Various functions of homologous pairing in somatic cells have been elucidated through high-throughput screens in the early 21st century.

Problems

There are severe repercussions when chromosomes do not segregate properly. Faulty segregation can lead to fertility problems, embryo death, birth defects, and cancer.

Though the mechanisms for pairing and adhering homologous chromosomes vary among organisms, proper functioning of those mechanisms is imperative in order for the final genetic material to be sorted correctly.

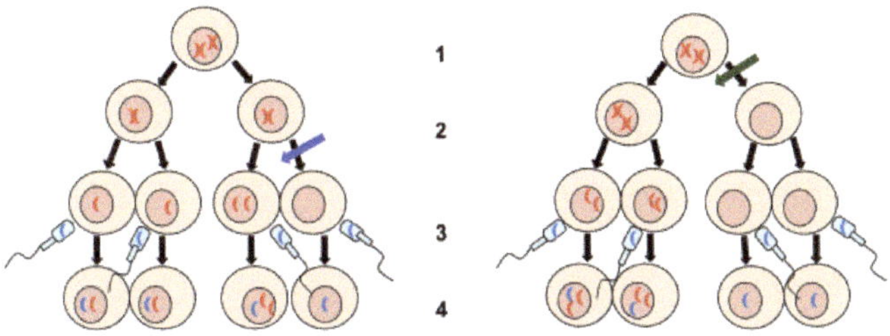

1. Meiosis I 2. Meiosis II 3. Fertilization 4. Zygote Nondisjunction is when chromosomes fail to separate normally resulting in a gain or loss of chromosomes. In the left image the blue arrow indicates nondisjunction taking place during meiosis II. In the right image the green arrow is indicating nondisjunction taking place during meiosis I.

Nondisjunction

Proper homologous chromosome separation in meiosis I is crucial for sister chromatid separation in meiosis II. A failure to separate properly is known as nondisjunction. There are two main types of nondisjunction that occur: trisomy and monosomy. Trisomy is caused by the presence of one additional chromosomes in the zygote as compared to the normal number, and monosomy is characterized by the presence of a one less chromosome in the zygote as compared to the normal number. If this uneven division occurs in meiosis I, then none of the daughter cells will have proper chromosomal distribution and severe effects can ensue, including Down's syndrome. Unequal division can also occur during the second meiotic division. Nondisjunction which occurs at this stage can result in normal daughter cells and deformed cells.

Embryonic Death

If unequal genetic crossover occurs within the gametes involved in meiosis, then the resulting zygote may suffer unviability. This is caused by events such as trisomy and monosomy because the cells are not able to develop normally with these characteristics. In other terms, one cell gains the proper amount of genetic material while the other cell lacks that material. Consequently, these deformed cells may cause a spontaneous abortion because the zygote does not develop correctly.

Other Uses

While the main function of homologous chromosomes is their use in nuclear division, they are also used in repairing double-strand breaks of DNA. These double-stranded breaks typically occur in DNA that serve as template strands for DNA replication, and they are the result of mutation, replication errors or malfunctioning DNA. Homologous chromosomes can repair this damage by aligning themselves with chromosomes of the same genetic sequence. Once the base pairs have been matched and oriented correctly between the two strands, the homologous chromosomes perform a process that is very similar to recombination, or crossing over as seen in meiosis. Part of the intact DNA sequence overlaps with that of the damaged chromosome's sequence. Replication proteins and complexes are then recruited to the site of damage, allowing for repair and proper replication to occur. Through this functioning, double-strand breaks can be repaired and DNA can function normally.

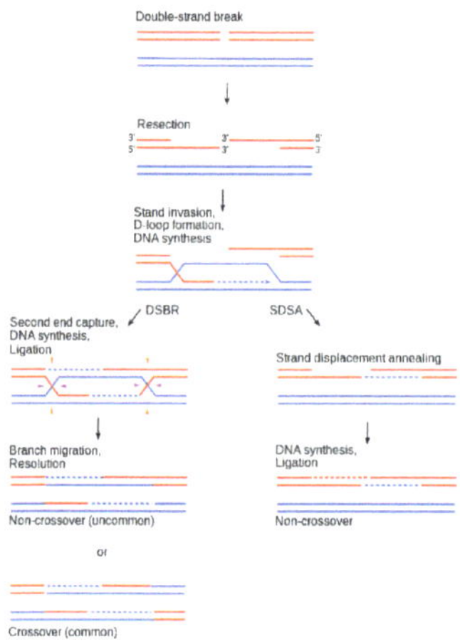

Diagram of the general process for double-stranded break repair as well as synthesis-dependent strand annealing.

Relevant Research

Current and future research on the subject of homologous chromosome is heavily focused on the roles of various proteins during recombination or during DNA repair. In a recently published article by Pezza et al. the protein known as HOP2 is responsible for both homologous chromosome synapsis as well as double-strand break repair via homologous recombination. The deletion of HOP2 in mice has large repercussions in meiosis. Other current studies focus on specific proteins involved in homologous recombination as well.

There is ongoing research concerning the ability of homologous chromosomes to repair double-strand DNA breaks. Researchers are investigating the possibility of exploiting this capability for regenerative medicine. This medicine could be very prevalent in relation to cancer, as DNA damage is thought to be contributor to carcinogenesis. Manipulating the repair function of homologous chromosomes might allow for bettering a cell's damage response system. While research has not yet confirmed the effectiveness of such treatment, it may become a useful therapy for cancer.

Sister Chromatids

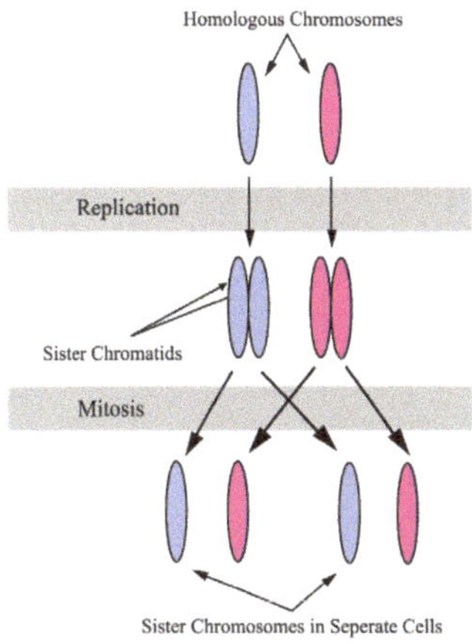

The paternal (blue) chromosomal and the maternal (pink) chromosome are homologous chromosomes. Following chromosomal replication, the blue chromosome is composed of two identical sister chromatids and the pink chromosome is composed of two identical sister chromatids. In mitosis, the sister chromatids separate into the daughter cells, but are now referred to as chromosomes (rather than chromatids) much in the way that one child is not referred to as a single twin.

A sister chromatid refers to either of the two identical copies (chromatids) formed by the replication of a single chromosome, with both copies joined together by a common centromere. In other words, a sister chromatid may also be said as 'one-half' of the duplicated chromosome. A full set of sister chromatids is created during the synthesis (S) phase of interphase, when all the chromosomes in a cell are replicated. The two sister chromatids are separated from each other into two different cells during mitosis and during the second division of meiosis straight after fertilization during sexual reproduction.

Compare sister chromatids to homologous chromosomes, which are the two *different* copies of a chromosome that diploid organisms (like humans) inherit, one from each parent. Sister chromatids are by and large identical (since they carry the same alleles, also called variants or versions, of genes) because they derive from one original chromosome. An exception is towards the end of meiosis, after crossing over has occurred, because sections of each sister chromatid may have been exchanged with corresponding sections of the homologous chromatids with which they are paired during meiosis. Homologous chromosomes might or might not be the same as each other because they derive from different parents.

There is evidence that, in some species, sister chromatids are the preferred template for DNA repair. Sister chromatid cohesion is essential for the correct distribution of genetic information between daughter cells and the repair of damaged chromosomes. Defects in this process may lead to aneuploidy and cancer, especially when checkpoints fail to detect DNA damage or when incorrectly attached mitotic spindles don't function properly.

Mitosis

Mitotic recombination is primarily a result of DNA repair processes responding to spontaneous or induced damages. (Also reviewed in Bernstein and Bernstein, pp 220–221 Homologous recombinational repair during mitosis is largely limited to interaction between nearby sister chromatids that are present in a cell subsequent to DNA replication but prior to cell division. Due to the special nearby relationship they share, sister chromatids are not only preferred over distant homologous chromatids as substrates for recominational repair, but have the capacity to repair more DNA damage than do homologs.

Meiosis

Studies with the budding yeast *Saccharomyces cerevisiae* indicate that inter-sister recombination occurs frequently during meiosis, and up to one-third of all recombination events occur between sister chromatids.

Cohesin

Cohesin is a protein complex that regulates the separation of sister chromatids during cell division, either mitosis or meiosis. Cohesins hold sister chromatids together after DNA replication until anaphase when removal of cohesin leads to separation of sister chromatids.

Structure

Cohesin is a multi-subunit protein complex, made up of four core subunits: two SMC proteins (SMC1 and 3), an alpha-kleisin (orthologues of yeast Scc1), and an orthologue of the yeast Scc3 protein (e.g. STAG1-3 in humans and SA1-3 in mice).

Smc1 and Smc3 are members of the Structural Maintenance of Chromosomes (SMC) family. SMC proteins have two main structural characteristics: an ATP-binding cassette-like 'head' domain with ATPase activity (formed by the interaction of the N- and C- terminals) and a hinge domain that allows dimerization of SMCs. The head and the hinge domains are connected to each other via long anti-parallel coiled coils. The dimer is present in a V-shaped form, connected by the hinges. Upon ATP binding, the two head domains in the dimer bind to each other, forming a ring structure. ATP hydrolysis can therefore trigger opening and closing of the ring.

Scc1 and Scc3 bind the ATPase domains of Smc1 and Smc3 stabilizing the ring structure. The amino and carboxy terminus of Scc1 bind Smc1 and Smc3. Once Scc1 binds the SMC proteins, Scc3 can also associate by binding with the C-terminal region of Scc1. When Scc1 binds on both Smc1 and Smc3, the cohesin complex forms a closed ring structure. When it binds to only one of the SMC proteins, the complex forms an open ring. Early studies suggested various ways in which cohesin may entrap DNA, including as a monomer that holds both homologues together, and a "hand-cuff" model where two intertwining cohesin complexes each hold one sister chromatid. While some studies support the idea of a hand-cuff model, the model is inconsistent with a number of experimental observations, and is generally considered to entrap chromatin as a monomer.

The topology and structure of these subunits has been best characterized in budding yeast (Haering et al. 2002, 2004), but the sequence conservation of these proteins and biochemical and electron microscopic observations imply that cohesin complexes in other species are very similar in their structure, .

Function

The cohesin ring has many functions:

1. It is used to keep the sister chromatids connected with each other during metaphase ensuring that during mitosis (and meiosis), each sister chromatid segregates to opposite poles. Without cohesin, the cell would be unable to control sister chromatid segregation since there would be no way of ensuring whether the spindle fiber attached on each sister chromatid is from a different pole.
2. It facilitates spindle attachment onto chromosomes.
3. It facilitates DNA repair by recombination.
4. Recently many novel functions of cohesin have been discovered in many different cellular processes. Cohesin has been shown to be responsible for transcription regulation, DNA double strand break repair, chromosome condensation, pairing of homologous chromosomes during meiosis I, mono-orientation of sister kinetochores during meiosis I, non-homologous centromere coupling, chromosome architecture and rearrangement, DNA replication etc.

Dissociation of Sister Chromatid Cohesion

The anaphase promoting complex associated to Cdc20 (APC/C-cdc20) cleaves the securin (anaphase inhibitor), which holds the sister chromatids together. Securin is cleaved at anaphase, following APC/C-cdc20 mediated degradation, and it renders separase (a protease, inhibited by the association with securin) to cleave the kleisin subunit. An alpha-kleisin is associated with the cohesin complex, linking both SMC 3 and SMC 1 together, with the exact kleisin varying between mitosis and meiosis (Scc1 and Rec8 respectively), and its cleavage ultimately leads to the removal of cohesin from chromosomes.

Dissociation of sister chromatids cohesion defines anaphase onset, which establishes two sets of identical chromosomes at each pole of the cell (telophase). Then the two daughter cells separate, and a new round of the cell cycle freshly starts in each one, at the stage of G0. When cells are ready to divide, because cell size is big enough or because they receive the appropriate stimulus, they activate the mechanism to enter into the G1 stage of cell cycle, and they duplicate most organelles during S (synthesis) phase, including their centrosome. Therefore, when the cell division process will end, each daughter cell will receive a complete set of organelles. At the same time, during S phase all cells must duplicate their DNA very precisely, a process termed DNA replication. Once DNA replication has finished, in eukaryotes the DNA molecule is compacted and condensed, to form the mitotic chromosomes, each one constituted by two sister chromatids, which stay held together by the establishment of cohesion between them; each chromatid is a complete DNA molecule, attached via microtubules to one of the two centrosomes of the dividing cell, located at opposed poles of the cell. To avoid premature sister chromatid separation, the APC/C is maintained in an inactive state bound to different molecules, which are part of a complex mechanism termed the spindle assembly checkpoint.

Mechanism of Action

It is not clear how the cohesion ring links sister chromatids together. There are two possible scenarios:

- Cohesin subunits bind to each sister chromatid and form a bridge between the two.
- Since cohesin has a ring structure, it is able to encircle both sister chromatids.

Current evidence suggests that the second scenario is the most likely. Proteins that are essential for sister chromatid cohesion, such as Smc3 and Scc1, do not regulate the formation of covalent bonds between cohesin and DNA, indicating that DNA interaction is not sufficient for cohesion. In addition, disturbing the ring structure of cohesin through cleavage of Smc3 or Scc1 triggers premature sister chromatid segregation in vivo. This shows that the ring structure is important for cohesin's function.

Even though the ring hypothesis appears to be valid, there are still questions about the number of rings required to hold sister chromatids together. One possibility is that one ring surrounds the two chromatids. Another possibility involves the creation of a dimer where each ring surrounds one sister chromatid. The two rings are connected to each other through formation of a bridge that holds the two sister chromatids together.

The cohesion complex is established during the initial stages of S-phase. The complexes associate with chromosomes before DNA replication occurs. Once cells start replicating their DNA, cohesin rings close and link the sister chromatids together. Cohesin complexes must be present during S-phase in order for cohesion to take place. It is unclear, however, how cohesin is loaded on the chromosomes during G1. There are two proposed hypotheses so far:

- The ATPase domain of the SMC proteins interacts with DNA and this interaction initially mediates the loading of cohesin complexes on chromosomes.
- Several proteins aid in the loading process. For example, Scc2 and Scc4 are both required for cohesin to load in budding yeast.

Localization of Cohesin Rings

Cohesin binding along the chromosomal DNA is considered to be dynamic and its location changes based on gene transcription, specific DNA sequence and presence of chromosome-associated proteins. There are three possible scenarios:

- Cohesin location is influenced by the orientation of neighboring genes and it is most frequently located in areas of convergent transcription. Gene orientation depends on the direction of transcription and can be of three types: head-to-head, head-to-tail and tail-to-tail. The tail-to-tail configuration results in the convergence of transcription machinery. One hypothesis states that the RNA polymerase "pushes" cohesin along the DNA, causing them to move towards the direction of the RNA polymerases. Changing the transcription pattern of genes changes the location of cohesin indicating that localization of cohesin may depend on transcription.
- A few cohesin rings are found in chromosome arms that have AT-rich DNA sequences indicating that DNA sequence may be an independent factor of cohesin binding.
- Cohesin rings, especially in budding yeast, are also located in the region surrounding the centromere. Two hypotheses may explain this: the presence of repetitive heterochromatic DNA in centromeres and the presence of chromosome-associated proteins. For example, Schizosaccharomyces pombe have multiple copies of specific heterochromatic DNA whose involvement in cohesion binding has been proven. Budding yeast lacks repetitive sequences and, therefore, requires a different mechanism for cohesion binding. Evidence suggests that binding of cohesin to the budding yeast centromere region depends on chromosome-associated proteins of the kinetochore that mediate cohesion association to pericentric regions (the kinetochore is an enhancer of pericentric cohesin binding).

Evolution

Cohesin structure and function has been conserved in evolution. The SMC proteins are found in prokaryotes and have been conserved through evolution. The coiled coils of SMC1 and SMC3 are conserved with an amino acid divergence of less than 0.5%.

Name	Saccharomyces cerevisiae	Schizosaccharomyces pombe	Drosophila	Vertebrates
Smc1	Smc1	Psm1	DmSmc1	Smc1
Smc3	Smc3	Psm3	DmSmc3	Smc3
Scc1	Mcd1/Pds3	Rad21	DmRad21	Rad21
Scc3	Scc3	Psc3	DmSA	SA1 and SA2

Clinical Significance

The term "cohesinopathy" has been used to describe conditions affecting the cohesin complex.

These conditions include:

- Cornelia de Lange Syndrome
- Roberts syndrome
- Warsaw breakage syndrome
- Many types of malignancies

Sister Chromatid Exchange

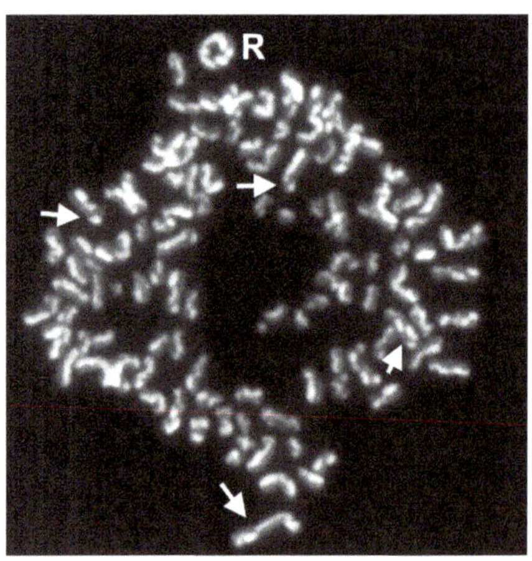

Metaphase spread of a cell line showing a ring chromosome (R) and several sister chromatid exchanges (SCEs), some of which are indicated by arrows.

Sister chromatid exchange (SCE) is the exchange of genetic material between two identical sister chromatids.

It was first discovered by using the Giemsa staining method on one chromatid belonging to the sister chromatid complex before anaphase in mitosis. The staining revealed that few segments were passed to the sister chromatid which were not dyed. The Giemsa staining was able to stain due to the presence of bromodeoxyuridine analogous base which was introduced to the desired chromatid.

The reason for the (SCE) is not known but it is required and used as a mutagenic testing of many products. Four to five sister chromatid exchange per chromosome pair, per mitosis is in the normal distribution, 14-100 exchanges is not normal and presents a danger to the organism. SCE is elevated in pathologies including Bloom syndrome, having recombination rates ~10-100 times above normal, depending on cell type. Frequent SCEs may also be related to formation of tumors.

Sister chromatid exchange has also been observed more frequently in B51(+) Behçet's disease.

Mitosis

Mitotic recombination in the budding yeast *Saccharomyces cerevisiae* is primarily a result of DNA repair processes responding to spontaneous or induced damages that occur during vegetative growth.} (Also reviewed in Bernstein and Bernstein, pp 220–221). In order for yeast cells to repair damage by homologous recombination, there must be present, in the same nucleus, a second DNA molecule containing sequence homology with the region to be repaired. In a diploid cell in G1 phase of the cell cycle, such a molecule is present in the form of the homologous chromosome. However, in the G2 phase of the cell cycle (following DNA replication), a second homologous DNA molecule is also present: the sister chromatid. Evidence indicates that, due to the special nearby relationship they share, sister chromatids are not only preferred over distant homologous chromatids as substrates for recominational repair, but have the capacity to repair more DNA damage than do homologs.

Meiosis

The genomes of diploid organisms in natural populations are highly polymorphic for insertions and deletions. During meiosis double-strand breaks (DSBs) that form within such polymorphic regions must be repaired by inter-sister chromatid exchange, rather than by inter-homolog exchange. A molecular-level study of recombination during budding yeast meiosis has shown that recombination events initiated by DSBs in regions that lack corresponding sequences in the non-sister homolog are efficiently repaired by inter-sister chromatid recombination. This recombination occurs with the same timing as inter-homolog recombination, but with reduced (2- to 3-fold) yields of Holliday junction joint molecules. This study, and comparable evidence from other organisms (e.g. Peacock), indicates that inter-sister recombination occurs frequently during meiosis, and up to one-third of all recombination events occur between sister chromatids, although mainly by a pathway that does not involve Holliday junction intermediates.

References

- Griffiths JF, Gelbart WM, Lewontin RC, Wessler SR, Suzuki DT, Miller JH (2005). Introduction to Genetic Analysis. New York: W.H. Freeman and Co. pp. 34–40, 473–476, 626–629. ISBN 0-7167-4939-4.
- Klug, William; Michael Cummings; Charlotte Spencer; Michael Pallodino (2009). "Chromosome Mutations: Variation in chromosome number and arrangement". In Beth Wilbur. Concepts of Genetics (9 ed.). San Francisco, CA: Pearson Benjamin Cumming. pp. 213–214. ISBN 9780321540980.
- Pollard TD, Earnshaw WC, Lippincott-Schwartz J (2008). Cell Biology (2 ed.). Philadelphia: Saunders/Elsevier. pp. 815, 821–822. ISBN 1-4160-2255-4.
- Morgan DL (2007). The cell cycle: principles of control. London: Published by New Science Press in association with Oxford University Press. ISBN 0-87893-508-8.
- Bernstein, C; Bernstein, H (1991). Aging, Sex, and DNA Repair. San Diego.: Academic Press. ISBN 978-0120928606.
- Lodish HF (2013). Molecular cell biolog. New York: W.H. Freeman and Co. pp. 355, 891. ISBN 1-4292-3413-X.

9

Microtubule: An Integrated Study

Microtubules are an essential part of the cytoskeleton in eukaryotic cells and certain types of bacteria. This chapter studies microtubules, their functions and their intracellular organization. Also included are topics about the various types of microtubules like preprophase band, astral microtubules and phycoplast.

Microtubule

Microtubules are a component of the cytoskeleton, found throughout the cytoplasm. These tubular polymers of tubulin can grow as long as 50 micrometres and are highly dynamic. The outer diameter of a microtubule is about 24 nm while the inner diameter is about 12 nm. They are found in eukaryotic cells, as well as some bacteria, and are formed by the polymerization of a dimer of two globular proteins, alpha and beta tubulin.

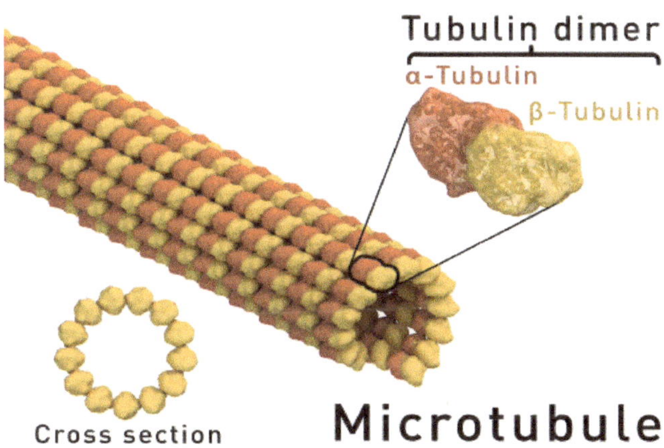

Structure of a microtubule. The ring shape depicts a microtubule in cross-section, showing the 13 protofilaments surrounding a hollow center.

Microtubules are very important in a number of cellular processes. They are involved in maintaining the structure of the cell and, together with microfilaments and intermediate filaments, they form the cytoskeleton. They also make up the internal structure of cilia and flagella.They provide platforms for intracellular transport and are involved in a variety of cellular processes, including the movement of secretory vesicles, organelles, and intracellular macromolecular assemblies. They are also involved in chromosome separation (mitosis and meiosis), and are the major constituents of mitotic spindles, which are used to pull apart eukaryotic chromosomes.

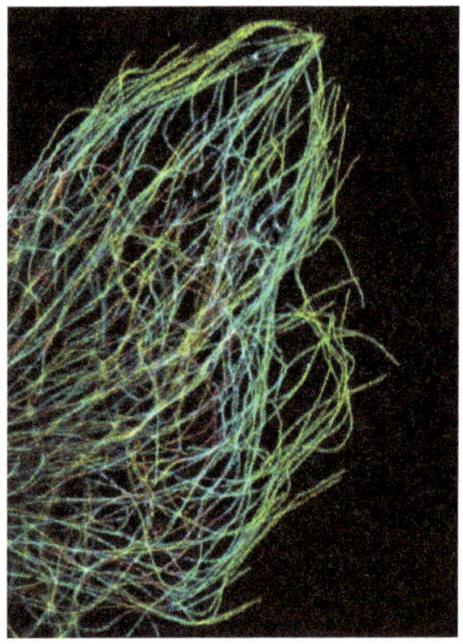

Just as our bodies rely on bones for structural support, our cells rely on a cellular skeleton. In addition to helping cells keep their shape, this cytoskeleton transports material within cells and coordinates cell division. Microtubules are shown here as thin strands

Microtubules are nucleated and organized by microtubule organizing centers (MTOCs), such as the centrosome found in the center of many animal cells or the basal bodies found in cilia and flagella, or the spindle pole bodies found in fungi.

There are many proteins that bind to microtubules, including the motor proteins kinesin and dynein, severing proteins like katanin, and other proteins important for regulating microtubule dynamics. Recently an actin-like protein has been found in a gram-positive bacterium *Bacillus thuringiensis*, which forms a microtubule-like structure and is involved in plasmid segregation.

History

Tubulin and microtubule-mediated processes, like cell locomotion, were seen by early microscopists, like Leeuwenhoek (1677). However, the fibrous nature of flagella and other structures were discovered two centuries later, with improved light microscopes, and confirmed in the 20th with electron microscope and biochemical studies.

Structure

In eukaryotes, microtubules are long, hollow cylinders made up of polymerised α- and β-tubulin dimers. The α and β-tubulin subunits are approximately 50% identical at the amino acid level, and each have a molecular weight of approximately 50 kDa.

These α/β-tubulin dimers polymerize end-to-end into linear protofilaments that associate laterally to form a single microtubule, which can then be extended by the addition of more α/β-tubulin dimers. Typically, microtubules are formed by the parallel association of thirteen protofilaments, although microtubules composed of fewer or more protofilaments have been observed *in vitro*.

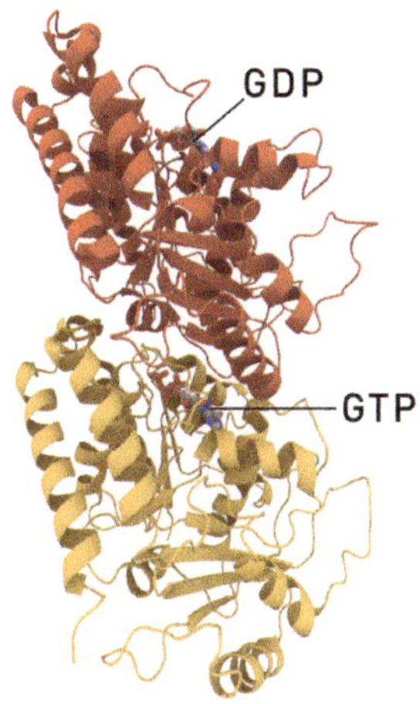

Cartoon representation of the structure of α(yellow)/β(red)-tubulin heterodimer, GTP and GDP.

Microtubules have a distinct polarity that is critical for their biological function. Tubulin polymerizes end to end, with the β-subunits of one tubulin dimer contacting the α-subunits of the next dimer. Therefore, in a protofilament, one end will have the α-subunits exposed while the other end will have the β-subunits exposed. These ends are designated the (−) and (+) ends, respectively. The protofilaments bundle parallel to one another with the same polarity, so, in a microtubule, there is one end, the (+) end, with only β-subunits exposed, while the other end, the (−) end, has only α-subunits exposed. While microtubule elongation can occur at both the (+) and (-) ends, it is significantly more rapid at the (+) end.

The lateral association of the protofilaments generates a pseudo-helical structure, with one turn of the helix containing 13 tubulin dimers, each from a different protofilament. In the most common "13-3" architecture, the 13th tubulin dimer interacts with the next tubulin dimer with a vertical offset of 3 tubulin monomers due to the helicity of the turn. There are other alternative architectures, such as 11-3, 12-3, 14-3, 15-4, or 16-4 , that have been detected at a much lower occurrence. Microtubules can also morph into other forms such as helical filaments, which are observed in protist organisms like foraminifera. There are two distinct types of interactions that can occur between the subunits of lateral protofilaments within the microtubule called the A-type and B-type lattices. In the A-type lattice, the lateral associations of protofilaments occur between adjacent α and β-tubulin subunits (i.e. an α-tubulin subunit from one protofilament interacts with a β-tubulin subunit from an adjacent protofilament). In the B-type lattice, the α and β-tubulin subunits from one protofilament interact with the α and β-tubulin subunits from an adjacent protofilament, respectively. Experimental studies have shown that the B-type lattice is the primary arrangement within microtubules. However, in most microtubules there is a seam in which tubulin subunits interact α-β.

Some species of *Prosthecobacter* also contain microtubules. The structure of these bacterial microtubules is similar to that of eukaryotic microtubules, consisting of a hollow tube of protofilaments assembled from heterodimers of bacterial tubulin A (BtubA) and bacterial tubulin B (BtubB). Both BtubA and BtubB share features of both α- and β-tubulin. Unlike eukaryotic microtubules, bacterial microtubules do not require chaperones to fold. In contrast to the 13 protofilaments of eukaryotic microtubules, bacterial microtubules comprise only five.

Intracellular Organization

Microtubules are part of a structural network (the cytoskeleton) within the cell's cytoplasm. Roles of the microtubule cytoskeleton include mechanical support, organization of the cytoplasm, transport, motility and chromosome segregation. A microtubule is capable of growing and shrinking in order to generate force, and there are motor proteins that allow organelles and other cellular components to be carried along a microtubule. This combination of roles makes microtubules important for organizing and moving intracellular constituents.

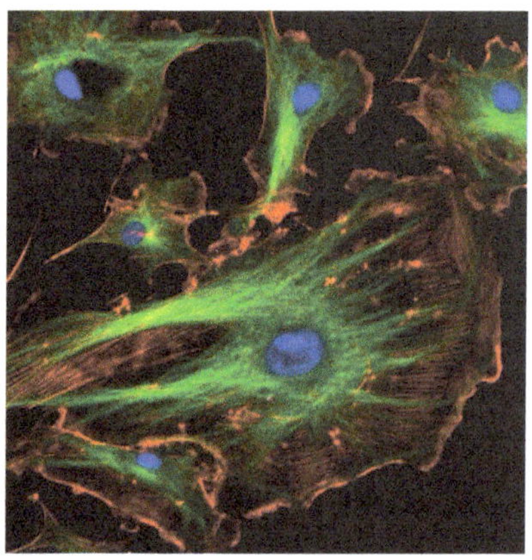

Components of the eukaryotic cytoskeleton. Actin filaments are shown in red, microtubules are in green, and the nuclei are in blue. The cystoskeleton provides the cell with an inner framework and enables it to move and change shape.

The organization of microtubules in the cell is cell-type specific. In epithelia, the minus-ends of the microtubule polymer are anchored near the site of cell-cell contact and organized along the apical-basal axis. After nucleation, the minus-ends are released and then re-anchored in the periphery by factors such as ninein and Nezha/PLEKHA7. In this manner, they can facilitate the transport of proteins, vesicles and organelles along the apical-basal axis of the cell. In fibroblasts and other mesenchymal cell-types, microtubules are anchored at the centrosome and radiate with their plus-ends outwards towards the cell periphery (as shown in the first figure). In these cells, the microtubules play important roles in cell migration. Moreover, the polarity of microtubules is acted upon by motor proteins, which organize many components of the cell, including the Endoplasmic Reticulum and the Golgi Apparatus.

Microtubule Polymerization

Nucleation

Microtubules are typically nucleated and organized by dedicated organelles called microtubule-organizing centres (MTOCs). Contained within the MTOC is another type of tubulin, γ-tubulin, which is distinct from the α- and β-subunits of the microtubules themselves. The γ-tubulin combines with several other associated proteins to form a lock washer-like structure known as the "γ-tubulin ring complex" (γ-TuRC). This complex acts as a template for α/β-tubulin dimers to begin polymerization; it acts as a cap of the (−) end while microtubule growth continues away from the MTOC in the (+) direction.

The centrosome is the primary MTOC of most cell types. However, microtubules can be nucleated from other sites as well. For example, cilia and flagella have MTOCs at their base termed basal bodies. In addition, work from the Kaverina group at Vanderbilt, as well as others, suggests that the Golgi apparatus can serve as an important platform for the nucleation of microtubules. Because nucleation from the centrosome is inherently symmetrical, Golgi-associated microtubule nucleation may allow the cell to establish asymmetry in the microtubule network. In recent studies, the Vale group at UCSF identified the protein complex augmin as a critical factor for centrosome-dependent, spindle-based microtubule generation. It that has been shown to interact with γ-TuRC and increase microtubule density around the mitotic spindle origin.

Some cell types, such as plant cells, do not contain well defined MTOCs. In these cells, microtubules are nucleated from discrete sites in the cytoplasm. Other cell types, such as trypanosomatid parasites, have a MTOC but it is permanently found at the base of a flagellum. Here, nucleation of microtubules for structural roles and for generation of the mitotic spindle is not from a canonical centriole-like MTOC.

Polymerization

Following the initial nucleation event, tubulin monomers must be added to the growing polymer. The process of adding or removing monomers depends on the concentration of αβ-tubulin dimers in solution in relation to the critical concentration, which is the steady state concentration of dimers at which there is no longer any net assembly or disassembly at the end of the microtubule. If the dimer concentration is greater than the critical concentration, the microtubule will polymerize and grow. If the concentration is less than the critical concentration, the length of the microtubule will decrease.

Microtubule Dynamics

Dynamic Instability

Dynamic instability refers to the coexistence of assembly and disassembly at the ends of a microtubule. The microtubule can dynamically switch between growing and shrinking phases in this region. Tubulin dimers can bind two molecules of GTP, one of which can be hydrolyzed subsequent to assembly. During polymerization, the tubulin dimers are in the GTP-bound state. The GTP bound to α-tubulin is stable and it plays a structural function in this bound state. However, the GTP bound to β-tubulin may be hydrolyzed to GDP shortly after assembly. The assembly properties of GDP-tubulin are different from those of GTP-tubulin, as GDP-tubulin is more prone to depolymerization. A GDP-bound tubulin subunit at the tip of a microtubule will tend to fall off, although a GDP-bound tubulin in the middle of a microtubule cannot spontaneously pop out of the polymer. Since tubulin adds onto the end of the microtubule in the GTP-bound state, a cap of GTP-bound tubulin is proposed to exist at the tip of the microtubule, protecting it from disassembly. When hydrolysis catches up to the tip of the microtubule, it begins a rapid depolymerization and shrinkage. This switch from growth to shrinking is called a catastrophe. GTP-bound tubulin can begin adding to the tip of the microtubule again, providing a new cap and protecting the microtubule from shrinking. This is referred to as "rescue".

Animation of the microtubule dynamic instability. Tubulin dimers bound to GTP (red) bind to the growing end of a microtubule and subsequently hydrolyze GTP into GDP (blue).

"Search and Capture" Model

In 1986, Marc Kirschner and Tim Mitchison proposed that microtubules use their dynamic properties of growth and shrinkage at their plus ends to probe the three dimensional space of the cell. Plus ends that encounter kinetochores or sites of polarity become captured and no longer display growth or shrinkage. In contrast to normal dynamic microtubules, which have a half-life of 5–10 minutes, the captured microtubules can last for hours. This idea is commonly known as the "search and capture" model. Indeed, work since then has largely validated this idea. At the kinetochore, a variety of complexes have been shown to capture microtubule (+)-ends. Moreover, a (+)-end capping activity for interphase microtubules has also been described. This later activity is mediated by formins, the adenomatous polyposis coli protein, and EB1, a protein that tracks along the growing plus ends of microtubules.

Regulation of Microtubule Dynamics

Post-translational Modifications

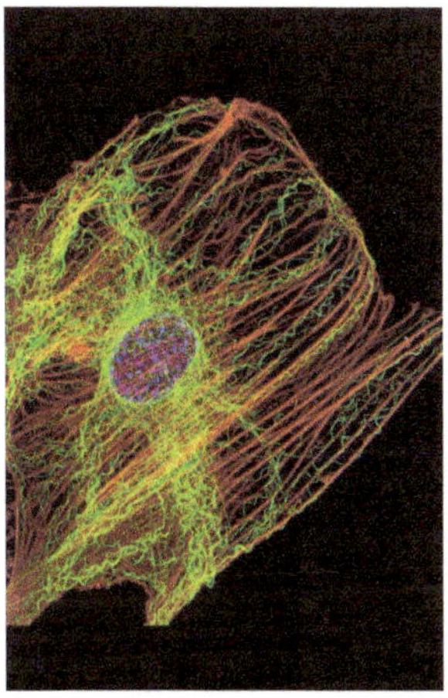

Image of a fibroblast cell containing fluorescently labeled actin (red) and microtubules (green).

Although most microtubules have a half-life of 5-10 min, certain microtubules can remain stable for hours. These stabilized microtubules accumulate post-translational modifications on their tubulin subunits by the action of microtubule-bound enzymes. However, once the microtubule depolymerizes, most of these modifications are rapidly reversed by soluble enzymes. Since most modification reactions are slow while their reverse reactions are rapid, modified tubulin is only detected on long-lived stable microtubules. Most of these modifications occur on the C-terminal region of alpha-tubulin. This region, which is rich in negatively charged glutamate, forms relativey unstructured tails that project out from the microtubule and form contacts with motors. Thus, it is believed that tubulin modifications regulate the interaction of motors with the microtubule. Since these stable modified microtubules are typically oriented towards the site of cell polarity in interphase cells, this subset of modified microtubules provide a specialized route that helps deliver vesicles to these polarized zones. These modifications include:

- Detyrosination: the removal of the C-terminal tyrosine from alpha-tubulin. This reaction exposes a glutamate at the new C-terminus. As a result, microtubules that accumulate this modification are often referred to as Glu-microtubules. Although the tubulin carboxypeptidase has yet to be identified, the tubulin—tyrosine ligase (TTL) is known.

- Delta2: the removal of the last two residues from the C-terminus of alpha-tubulin. Unlike detyrosination, this reaction is thought to be irreversible and has only been documented in neurons.
- Acetylation: the addition of an acetyl group to lysine 40 of alpha-tubulin. This modification occurs on a lysine that is accessible only from the inside of the microtubule, and it remains unclear how enzymes access the lysine residue. The nature of the tubulin acetyltransferase remains controversial; however, the reverse reaction is known to be catalyzed by HDAC6.
- Polyglutamylation: the addition of a glutamate polymer (typically 4-6 residues long) to the gamma-carboxyl group of any one of five glutamates found near the end of alpha-tubulin. Enzymes related to TTL add the initial branching glutamate (TTL4,5 and 7), while other enzymes that belong to the same family lengthen the polyglutamate chain (TTL6,11 and 13).
- Polyglycylation: the addition of a glycine polymer (2-10 residues long) to the gamma-carboxyl group of any one of five glutamates found near the end of beta-tubulin. TTL3 and 8 add the initial branching glycine, while TTL10 lengthens the polyglycine chain.

Tubulin is also known to be phosphorylated, ubiquitinated, sumoylated, and palmitoylated.

Tubulin-binding Drugs and Chemical Effects

A wide variety of drugs are able to bind to tubulin and modify its assembly properties. These drugs can have an effect at intracellular concentrations much lower than that of tubulin. This interference with microtubule dynamics can have the effect of stopping a cell's cell cycle and can lead to programmed cell death or apoptosis. However, there are data to suggest that interference of microtubule dynamics is insufficient to block the cells undergoing mitosis. These studies have demonstrated that suppression of dynamics occurs at concentrations lower than those needed to block mitosis. Suppression of microtubule dynamics by tubulin mutations or by drug treatment have been shown to inhibit cell migration. Both microtubule stabilizers and destabilizers can suppress microtubule dynamics.

The drugs that can alter microtubule dynamics include:

- The cancer-fighting taxane class of drugs (paclitaxel (taxol) and docetaxel) block dynamic instability by stabilizing GDP-bound tubulin in the microtubule. Thus, even when hydrolysis of GTP reaches the tip of the microtubule, there is no depolymerization and the microtubule does not shrink back.

- The epothilones, e.g. Ixabepilone, work in a similar way to the taxanes.
- Nocodazole, vincristine, and colchicine have the opposite effect, blocking the polymerization of tubulin into microtubules.
- Eribulin binds to the (+) growing end of the microtubules. Eribulin exerts its anticancer effects by triggering apoptosis of cancer cells following prolonged and irreversible mitotic blockade.

Expression of β3-tubulin has been reported to alter cellular responses to drug-induced suppression of microtubule dynamics. In general the dynamics are normally suppressed by low, subtoxic concentrations of microtubule drugs that also inhibit cell migration. However, incorporating β3-tubulin into microtubules increases the concentration of drug that is needed to suppress dynamics and inhibit cell migration. Thus, tumors that express β3-tubulin are not only resistant to the cytotoxic effects of microtubule targeted drugs, but also to their ability to suppress tumor metastasis. Moreover, expression of β3-tubulin also counteracts the ability of these drugs to inhibit angiogenesis which is normally another important facet of their action.

Microtubule polymers are extremely sensitive to various environmental effects. Very low levels of free calcium can destabilize microtubules and this prevented early researchers from studying the polymer in vitro. Cold temperatures also cause rapid depolymerization of microtubules. In contrast, heavy water promotes microtubule polymer stability.

Proteins that Interact with Microtubules

Microtubule-associated Proteins (MAPs)

MAPs have been shown to play a crucial role in the regulation of microtubule dynamics *in-vivo*. The rates of microtubule polymerization, depolymerization, and catastrophe vary depending on which microtubule-associated proteins (MAPs) are present. The originally identified MAPs from brain tissue can be classified into two groups based on their molecular weight. This first class comprises MAPs with a molecular weight below 55-62 kDa, and are called τ (tau) proteins. *In-vitro*, tau proteins have been shown to directly bind microtubules, promote nucleation and prevent disassembly, and to induce the formation of parallel arrays. Additionally, tau proteins have also been shown to stabilize microtubules in axons and have been implicated in Alzheimer's disease. The second class is composed of MAPs with a molecular weight of 200-1000 kDa, of which there are four known types: MAP-1, MAP-2, MAP-3 and MAP-4. MAP-1 proteins consists of a set of three different proteins: A, B and C. The C protein plays an important role in the retrograde transport of vesicles and is also known as cytoplasmic dynein. MAP-2 proteins are located in the dendrites and in the body of neurons, where they bind with other cytoskeletal filaments. The MAP-4 proteins are found in the majority of cells and stabilize microtubules. In addition to MAPs that have a stabilizing effect on microtubule structure, other MAPs can have a destabilizing effect either by cleaving or by inducing depolymerization of microtubules. Three proteins called katanin, spastin, and fidgetin have been observed to regulate the number and length of microtubules via their destabilizing activities.

Plus-end Tracking Proteins (+TIPs)

Plus end tracking proteins are MAP proteins which bind to the tips of growing microtubules and play an important role in regulating microtubule dynamics. For example, +TIPs have been observed to participate in the interactions of microtubules with chromosomes during mitosis. The first MAP to be identified as a +TIP was CLIP170 (cytoplasmic linker protein), which has been shown to play a role in microtubule depolymerization rescue events. Additional examples of +TIPs include EB1, EB2, EB3, p150Glued, Dynamitin, Lis1, CLIP115, CLASP1, and CLASP2.

Motor Proteins

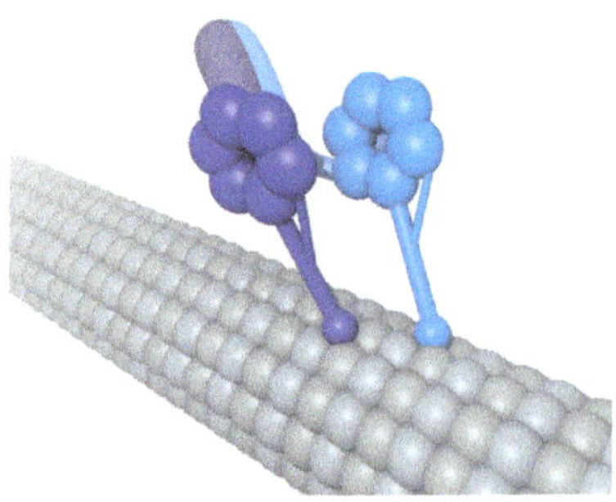

A cytoplasmic dynein motor bound to a microtubule.

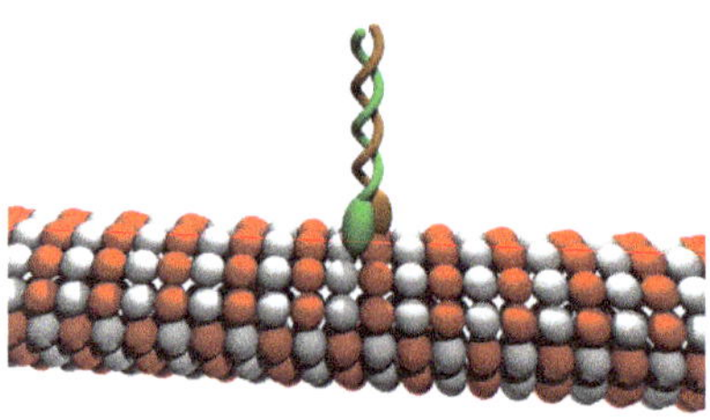

A kinesin molecule bound to a microtubule.

Microtubules can act as substrates for motor proteins that are involved in important cellular functions such as vesicle trafficking and cell division. Unlike other microtubule-associated proteins, motor proteins utilize the energy from ATP hydrolysis to generate mechanical work that moves the protein along the substrate. The major motor proteins that interact with microtubules are kinesin, which usually moves toward the (+) end of the microtubule, and dynein, which moves toward the (−) end.

- Dynein is composed of two identical heavy chains, which make up two large globular head domains, and a variable number of intermediate and light chains. Dynein-mediated transport takes place from the (+) end towards the (-) end of the microtubule. ATP hydrolysis occurs in the globular head domains, which share similarities with the AAA+ (ATPase associated with various cellular activities) protein family. ATP hydolysis in these domains is coupled to movement along the microtubule via the microtubule-binding domains. Dynein transports vesicles and organelles throughout the cytoplasm. In order to do this, dynein molecules bind organelle membranes via a protein complex that contains a number of elements including dynactin.
- Kinesin has a similar structure to dynein. Kinesin is involved in the transport of a variety of intracellular cargoes, including vesicles, organelles, protein complexes, and mRNAs toward the microtubule's (+) end.

Some viruses (including retroviruses, herpesviruses, parvoviruses, and adenoviruses) that require access to the nucleus to replicate their genomes attach to motor proteins.

Functions

Cell Migration

Microtubule plus ends are often localized to particular structures. In polarized interphase cells, microtubules are disproportionately oriented from the MTOC toward the site of polarity, such as the leading edge of migrating fibroblasts. This configuration is thought to help deliver microtubule-bound vesicles from the Golgi to the site of polarity.

Dynamic instability of microtubules is also required for the migration of most mammalian cells that crawl. Dynamic microtubules regulate the levels of key G-proteins such as RhoA and Rac1, which regulate cell contractility and cell spreading. Dynamic microtubules are also required to trigger focal adhesion disassembly, which is necessary for migration.

Mitosis

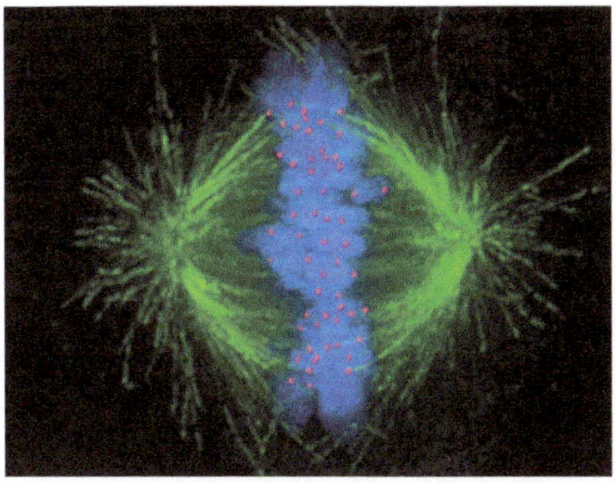

Image of the mitotic spindle in a human cell showing microtubules in green, (chromosomes (DNA) in blue, and kinetochores in red).

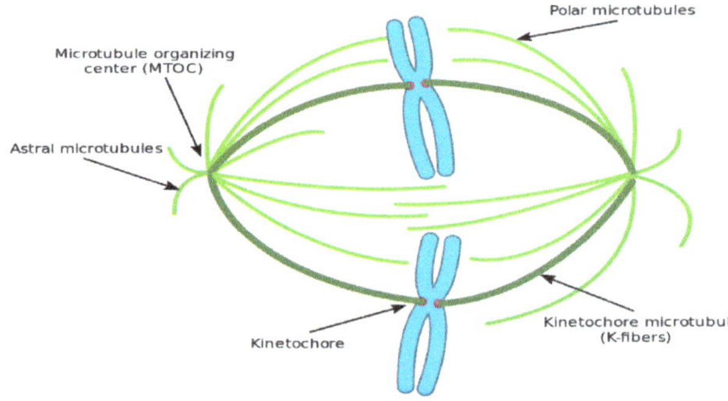

This diagram depicts the organization of a typical mitotic spindle found in animal cells. Chromosomes are attached to kinetochore microtubules via a multiprotein complex called the kinetochore. Polar microtubules interdigitate at the spindle midzone and push the spindle poles apart via motor proteins. Astral microtubules anchor the spindle poles to the cell membrane. Microtubule polymerization is nucleated at the microtubule organizing center.

A notable structure composed largely of microtubules is the mitotic spindle, used by eukaryotic cells to segregate their chromosomes during cell division. The mitotic spindle includes the spindle microtubules, microtubule-associated proteins (MAPs), and the MTOC. The microtubules originate in the MTOC and fan out into the cell; each cell has two MTOCs, as shown in the diagram.

The process of mitosis is facilitated by three main subgroups of microtubules, known as astral, polar, and kinetochore microtubules. An astral microtubule is a microtubule originating from the MTOC that does not connect to a chromosome. Astral microtubules instead interact with the cytoskeleton near the cell membrane and function in concert with specialized dynein motors. Dynein motors pull the MTOC toward the cell membrane, thus assisting in correct positioning and orientation of the entire apparatus.

Kinetochore microtubules directly connect to the chromosomes, at the kinetochores. To clarify the terminology, each chromosome has two chromatids, and each chromatid has a kinetochore. The two kinetochores associated with a region of the chromosome called the centromere. The polar microtubules from one MTOC intertwine with the microtubules from the other MTOC; motor proteins make them push against each other and assist in the separation of the chromosomes to the two daughter cells.

Cell division in a typical eukaryote finishes with the generation of a final cytoplasmic bridge between the two daughter cells termed the midbody. This structure is built of microtubules that originally made up part of the mitotic spindle.

Cilia and Flagella

Microtubules have a major structural role in eukaryotic cilia and flagella. Cilia and flagella always extend directly from a MTOC, in this case termed the basal body. The action of the dynein motor proteins on the various microtubule strands that run along a cilium or flagellum allows the organelle to bend and generate force for swimming, moving extracellular material, and other roles. Prokaryotes possess tubulin-like proteins including FtsZ. However, prokaryotic flagella are entirely different in structure from eukaryotic flagella and do not contain microtubule-based structures.

Development

The cytoskeleton formed by microtubules is essential to the morphogenetic process of an organism's development. For example, a network of polarized microtubules is required within the oocyte of *Drosophila melanogaster* during its embryogenesis in order to establish the axis of the egg. Signals sent between the follicular cells and the oocyte (such as factors similar to epidermal growth factor) cause the reorganization of the microtubules so that their (-) ends are located in the lower part of the oocyte, polarizing the structure and leading to the appearance of an anterior-posterior axis. This involvement in the body's architecture is also seen in mammals.

Another area where microtubules are essential is the formation of the nervous system in higher vertebrates, where tubulin's dynamics and those of the associated proteins (such as the MAPs) is finely controlled during the development of the nervous system.

Gene Regulation

The cellular cytoskeleton is a dynamic system that functions on many different levels: In addition to giving the cell a particular form and supporting the transport of vesicles and organelles, it can also influence gene expression. However, the signal transduction mechanisms involved in this communication are little understood. Notwithstanding this, the relationship between the drug-mediated depolymerization of microtubules and the specific expression of transcription factors has been described, which has provided information on the differential expression of the genes depending on the presence of these factors. This communication between the cytoskeleton and the regulation of the cellular response is also related to the action of growth factors: for example, this relation exists for connective tissue growth factor.

Preprophase Band

The preprophase band is a microtubule array found in plant cells that are about to undergo cell division and enter the preprophase stage of the plant cell cycle. Besides the phragmosome, it is the first microscopically visible sign that a plant cell is about to enter mitosis. The preprophase band was first observed and described by Jeremy Pickett-Heaps and Donald Northcote at Cambridge University in 1966.

Just before mitosis starts, the preprophase band forms as a dense band of microtubules around the phragmosome and the future division plane just below the plasma membrane. It encircles the nucleus at the equatorial plane of the future mitotic spindle when dividing cells enter the G2 phase of the cell cycle after DNA replication is complete. The preprophase band consists mainly of microtubules and microfilaments (actin) and is generally 2-3 μm wide. When stained with fluorescent markers, it can be seen as two bright spots close to the cell wall on either side of the nucleus.

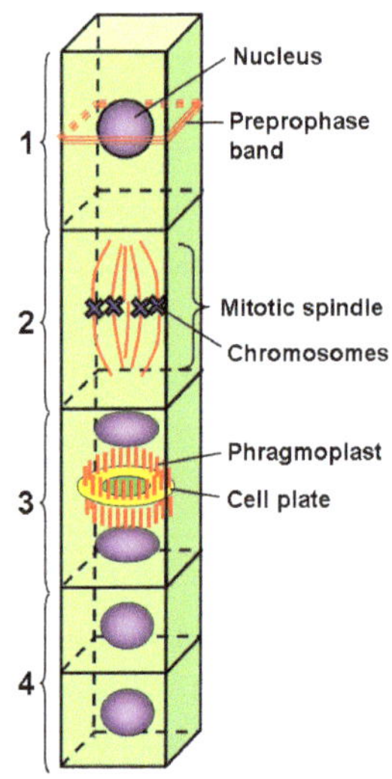

The preprophase band predicts the cell division plane: 1) Preprophase band formation during preprophase. 2) Metaphase spindle orients with the equator along the plane marked by preprophase band. 3) Phragmoplast and cell plate form along the plane marked by preprophase band. 4) The new cell wall of the daughter cells connects with the parent cell wall along the line of the former preprophase band location.

Plant cells lack centrosomes as microtubule organizing centers. Instead, the microtubules of the mitotic spindle aggregate on the nuclear surface and are reoriented to form the spindle at the end of prophase. The preprophase band also functions in properly orienting the mitotic spindle, and contributes to efficient spindle formation during prometaphase

The preprophase band disappears as soon as the nuclear envelope breaks down and the mitotic spindle forms, leaving behind an actin-depleted zone. However, its position marks the future fusion sites for the new cell plate with the existing cell wall during telophase. When mitosis is completed, the cell plate and new cell wall form starting from the center along the plane occupied by the phragmosome. The cell plate grows outwards until it fuses with the cell wall of the dividing cell at exactly the spots predicted by the position of the preprophase band.

Astral Microtubules

Astral microtubules are a subpopulation of microtubules, which only exist during and immediately before mitosis. They are defined as any microtubule originating from the centrosome which does not connect to a kinetochore. Astral microtubules develop in the actin skeleton and interact with the cell cortex to aid in spindle orientation. They are organized into radial arrays around the centrosomes. The turn-over rate of this population of microtubules is higher than any other population.

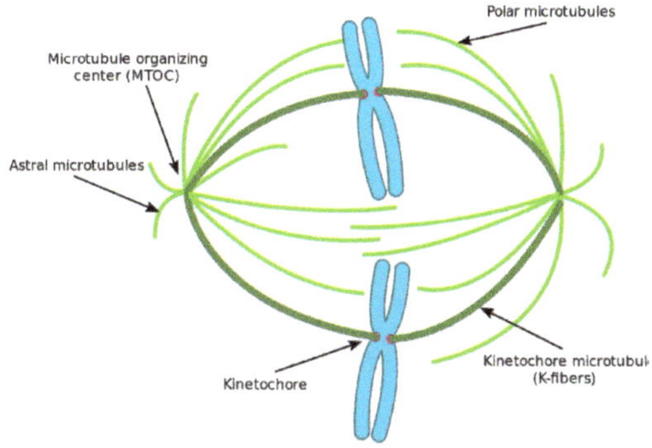

This diagram depicts the organization of a typical mitotic spindle found in animal cells. Chromosomes are attached to kinetochore microtubules via a multiprotein complex called the kinetochore. Polar microtubules interdigitate at the spindle midzone and push the spindle poles apart via motor proteins. Astral microtubules anchor the spindle poles to the cell membrane. Microtubule polymerization is nucleated at the microtubule organizing center.

The role of astral microtubules is assisted by dyneins specific to this role. These dyneins have their light chains (static portion) attached to the cell membrane, and their globular parts (dynamic portions) attached to the microtubules. The globular chains attempt to move towards the centrosome, but as they are bound to the cell membrane, this results in pulling the centrosomes towards the membrane, thus assisting cytokinesis.

Astral microtubules are not required for the progression of mitosis, but they are required to ensure the fidelity of the process. The function of astral microtubules can be generally considered as determination of cell geometry. They are absolutely required for correct positioning and orientation of the mitotic spindle apparatus, and are thus involved in determining the cell division site based on the geometry and polarity of the cells.

The maintenance of astral microtubules is dependent on the integrity of centrosome. It is also dependent on several microtubule-associated proteins such as EB1 and adenomatous polyposis coli (APC).

Phycoplast

The phycoplast is a microtubule structure observed during cytokinesis in members of the *Chlorophyceae*, the largest class of green algae.

Cytokinesis in green algae occurs via a diverse range of mechanisms, including cleavage furrows in some algae and cell plates in others. Some green algae of the class *Charophyceae* use phragmoplasts similar to those in embryophytes to organize and guide the growing cell plate. In these algae, the microtubules of the telophase spindle give rise to the phragmoplast and are oriented perpendicular to the plane of cell division and the forming cell plate. The growth of the cell plate eventually disrupts the telophase spindle.

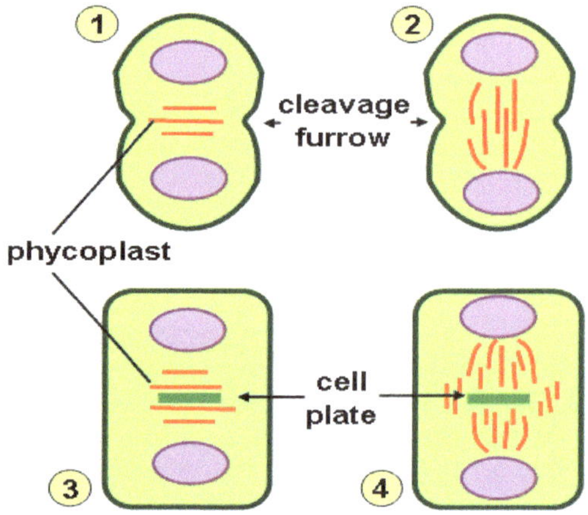

Schematic representation of types of cytokinesis in the green algae: 1) Phycoplast formation with cleavage furrow (e.g. *Chlamydomonas*); 2) Cleavage furrow and persistent telophase spindle (e.g. *Klebsormidium*); 3) Phycoplast and cell plate formation (e.g. *Fritschiella*); 4) Persistent telophase spindle/ phragmoplast with cell plate formation (e.g. *Coleochaete*)

In the *Chlorophyceae*, the most common form of cell division occurs via a phycoplast. In these algae, the spindle collapses and a new system of microtubules forms that is oriented in parallel to the plane of cell division. This phycoplast can be observed in algae undergoing cytokinesis via cleavage furrow (case 1 in picture) as well as algae utilizing a cell plate (case 3 in picture). The phycoplast may play a role in assuring that the plane of cell division will pass between the two daughter nuclei. Typically, these algae undergo "closed" mitosis where the nuclear envelope persists throughout mitosis.

References

- Karp, Gerald (2005). Cell and Molecular Biology: Concepts and Experiments. USA: John Wiley & Sons. p. 355. ISBN 0-471-46580-1.
- P.H. Raven, R.F. Evert, S.E. Eichhorn (2005): Biology of Plants, 7th Edition, W.H. Freeman and Company Publishers, New York, ISBN 0-7167-1007-2
- Ganguly, A; Yang, H; Cabral, F (2011). "Class III β-tubulin counteracts the ability of paclitaxel to inhibit cell migration.". Oncotarget. 2 (5): 368–377. PMC 3248193. PMID 21576762.

Permissions

All chapters in this book are published with permission under the Creative Commons Attribution Share Alike License or equivalent. Every chapter published in this book has been scrutinized by our experts. Their significance has been extensively debated. The topics covered herein carry significant information for a comprehensive understanding. They may even be implemented as practical applications or may be referred to as a beginning point for further studies.

We would like to thank the editorial team for lending their expertise to make the book truly unique. They have played a crucial role in the development of this book. Without their invaluable contributions this book wouldn't have been possible. They have made vital efforts to compile up to date information on the varied aspects of this subject to make this book a valuable addition to the collection of many professionals and students.

This book was conceptualized with the vision of imparting up-to-date and integrated information in this field. To ensure the same, a matchless editorial board was set up. Every individual on the board went through rigorous rounds of assessment to prove their worth. After which they invested a large part of their time researching and compiling the most relevant data for our readers.

The editorial board has been involved in producing this book since its inception. They have spent rigorous hours researching and exploring the diverse topics which have resulted in the successful publishing of this book. They have passed on their knowledge of decades through this book. To expedite this challenging task, the publisher supported the team at every step. A small team of assistant editors was also appointed to further simplify the editing procedure and attain best results for the readers.

Apart from the editorial board, the designing team has also invested a significant amount of their time in understanding the subject and creating the most relevant covers. They scrutinized every image to scout for the most suitable representation of the subject and create an appropriate cover for the book.

The publishing team has been an ardent support to the editorial, designing and production team. Their endless efforts to recruit the best for this project, has resulted in the accomplishment of this book. They are a veteran in the field of academics and their pool of knowledge is as vast as their experience in printing. Their expertise and guidance has proved useful at every step. Their uncompromising quality standards have made this book an exceptional effort. Their encouragement from time to time has been an inspiration for everyone.

The publisher and the editorial board hope that this book will prove to be a valuable piece of knowledge for students, practitioners and scholars across the globe.

Index